Analysis
of Variance

Analysis of Variance

LISA CUSTER
DANIEL R. McCARVILLE
MIKEL J. HARRY
JACK PRINS

ADDISON-WESLEY PUBLISHING COMPANY, INC.
Reading, Massachusetts   Menlo Park, California   New York
Don Mills, Ontario   Wokingham, England   Amsterdam
Bonn   Paris   Milan   Madrid   Sydney   Singapore
Tokyo   Seoul   Taipei   Mexico City   San Juan

Many of the designations used by manufacturers and sellers to distinguish their products are claimed as trademarks. Where those designations appear in this book and Addison-Wesley was aware of a trademark claim, the designations have been printed with initial capital letters.

The publisher offers discounts on this book when ordered in quantity for special sales.

For more information, please contact:
Corporate & Professional Publishing Group
Addison-Wesley Publishing Company
One Jacob Way
Reading, Massachusetts 01867

Library of Congress Cataloging-in-Publication Data

Analysis of variance / Lisa Custer . . . [et al.].—Motorola Univ.
    Press partnership ed.
        p.   cm.—(Six Sigma Research Institute series)
    Includes index.
    ISBN 0-201-63402-3
    1. Analysis of variance.   I. Custer, Lisa.   II. Series.
QA279.A5244   1993
519.5′38—dc20   92-33531
                                                             CIP

Copyright © 1993 by Motorola, Inc.

ISBN 0-201-63402-3

1 2 3 4 5 6 7 8 9 AL 96959493

First Printing: April 1993

Motorola University Press
Six Sigma Research Institute Publications

The mission of Motorola's Six Sigma Research Institute is to research and develop the theoretical framework and supporting tools necessary to achieve Six Sigma Quality. Six Sigma Quality can be thought of as a measurement of perfection, since it means that for every million opportunities there are to create a defect, only 3.4 defects will be produced.

The three predominant sources of product variation—inadequate design margin, insufficient process control, and unstable or varying materials and components—must be simultaneously improved to achieve Six Sigma Quality. Improvements in these same areas are also applied to service and other nonmanufactured products.

As part of its mission, the Six Sigma Research Institute authors are creating an entire set of analytical tools and case studies that can be used to achieve robust product and service designs. Motorola University Press books document these tools and studies. This series of books also underscores the importance of sharing knowledge and skills with Motorola's technical and management community, Motorola's partners, and preferred suppliers. The list of publications below reflects the need to improve quality across all areas of an organization.

Six Sigma Tool Publications

These tool publications are designed for engineers and statisticians with a working knowledge of basic statistics who need a quick reference to key techniques. Some tool publications may require readers to have taken one or more advanced statistics course.

Indices of Capability	*Data Display*
Tests for Differences	*Metrology*
Design of Experiments	*Correlation*
Basic Statistics	*Data Transformations*
Analysis of Variance	*Decision Theory*
Probability	*Mathematical Statistics*

Probability Distributions *Optimization Methods*
Process Control Methods *Regression Analysis*
Nonparametric Statistics *Sample Size Estimation*
Grouped Data

Six Sigma Case Study Publications

These publications cover actual cases of the application of Six Sigma concepts and techniques. The cases cross technical and nontechnical functional areas, including software engineering, silicon wafer fabrication, facilities management, manufacturing, and supplier selection. The titles listed below represent a sampling of the initial case study publications in this series. Please call 1-800-238-9682 for a brochure describing all of the titles that are currently available in the Six Sigma Research Institute series.

Productivity Analysis and Optimization

Optimization of a Liquid Crystal-Display to a Heat-Seal Connector Process

Characterizing and Measuring Mechanical Design and Productivity

Software Design Analysis and Optimization

Engineering Quality Software

Mechanical Design Tolerance Analysis

A Six Sigma Approach to the Process Design of Printed Circuit Board Screws

Desktop Computer Keyboard Failure Analysis

Digital Design Analysis and Optimization

Development of a Modular Design Methodology to Facilitate Design Reuse

Six Sigma Design of a Wideband Digital Communication System

Process Design Analysis and Optimization

Designing a New Factory with Manufacturing Simulation and Planned Experimentation

Methodology Development for Electronic Database Migration

Process Characterization and Optimization

Process Characterization and Optimization

Analog Design Analysis and Optimization

*Reduction of JFET Parameter Drift in IC Operational
Amplifiers Using Statistical Process Characterization
The Generation of System Moments for Improving Product
Quality in Advanced IC Manufacture*

Radio Frequency Analysis and Optimization

*Radio Frequency / Microwave Robust Design Techniques
Circuit Simulation and Optimization Using Matrix
Analysis*

Supplier Selection, Certification, and Control

*Supplier Mold Compound Preform Manufacturing Process
Characterization
Supplier Selection and Improvement*

Service Design, Analysis, and Optimization

*Application of Six Sigma Concepts to a Facilities
Organization
A Methodology for Measurement of Publications Quality*

Statistical Process Control

*Projecting Field Fallout Rates with Manufacturing Data
Statistical Process Control for a Multifarious System*

Six Sigma Software Packages

These highly interactive software packages provide
hands-on experience for statisticians and engineers
already familiar with basic statistical techniques. All
software is accompanied by a comprehensive manual
and instructions for use.

*Six Sigma Quality Control Charts (SSQC)
Six Sigma Metrics (SSMET)
Black Box Simulator*

Foreword

Motorola has set a new quality goal of Six Sigma and Beyond for the 1990s and the new millennium. This translates to an initial error rate of 3.4 parts per million, followed by a ten-fold improvement every two years. To accelerate the attainment of this goal, Motorola established the Six Sigma Research Institute (SSRI) in 1990. Its mission was clearly defined—to pioneer new tools, methods, and approaches in the quest for Six Sigma and Beyond.

World-class corporations joined the quest—Asea Brown Boveri (ABB), Digital Equipment Corporation (DEC), Eastman Kodak, International Business Machines (IBM), and Texas Instruments. As full partners of SSRI, these corporations defined two fundamental objectives. The first objective is to develop advanced engineering, manufacturing, and service optimization methods. The second objective is to devise a series of interactive educational resources that will decrease knowledge transfer cycle-time and concurrently increase training effectivity on the job.

In response to these needs, SSRI has developed many unique resources, one of which you are now holding. The Six Sigma Series has taken the efforts of many people having a wide variety of talents and experiences. Many thanks and much gratitude to:

- Motorola management for their vision and support,

- SSRI Partners for their trust and confidence,

- Our authors who have set new standards for excellence, and

- Motorola University Press and Addison-Wesley Publishing Company for the end product—a series of world-class publications.

  Special acknowledgment and thanks are given to the following individuals:

- Dr. Jack Prins, who helped find and select the authors;

- Mr. Doug Mader, who critiqued the authors' work;

- Ms. Sandy Carsello, Ms. Betty Crofton, and Dr. Peter Kusel, who communicated with the authors and processed their manuscripts;

- Mr. Kenneth H. Paterson, Publisher, and Ms. Susan Schneider, Associate Publisher of Motorola University Press, who worked with Addison-Wesley on the publishing of these works; and

- Mr. Bill Wiggenhorn, President of Motorola University, for his enthusiastic encouragement and support.

Mikel J. Harry, Ph.D.
Director and Senior Member of Technical Staff
Six Sigma Research Institute

Contents

About the Authors

Lisa Custer joined Motorola's Semiconductor Products Sector in Phoenix, Arizona in 1988. She is currently the quality manager working in the Hybrid Power Module Operations. Her responsibilities include the implementation of quality programs within the product design and manufacturing organizations of Hybrid Power Module Operations and programs to promote supplier partnerships. Custer earned a B.S. degree in mechanical engineering from Georgia Institute of Technology in 1988, an M.S. degree in engineering from Arizona State University in 1991, and is currently pursuing her Ph.D. in industrial engineering, with an emphasis in applied statistics and quality control, at Arizona State University.

Daniel R. McCarville has been with Motorola since 1984, and is the manufacturing engineering manager for Sensor Products organization of Motorola's Semiconductor Products Section in Phoenix, Arizona. His current responsibilities include the implementation of statistical techniques in the manufacturing areas of Sensor Products. He received a B.S. degree in mechanical engineering at the University of Arizona in 1984. McCarville is currently working on his M.S. and Ph.D. degrees in industrial engineering with an emphasis in applied statistics and quality control at Arizona State University.

Mikel J. Harry joined Motorola in 1985 and is currently corporate director of the Six Sigma Research Institute and senior member of Technical Staff within Motorola University. He is also an Associate Member of Motorola's Science Advisory Board (SABA) and the Scientific and Technical Society at the Motorola Government Electronics Group. He received his B.S. degree in industrial technology (electronics) from Ball State University in 1973. His M.A. degree was awarded in 1981, also from Ball State University. In 1984, he received his Ph.D. degree from Arizona State University. In his current assignment, Dr. Harry is responsible for the research, development, and subsequent documentation of advanced statistical engineering models and methods. For pioneering work, he has received several major engineering awards from Motorola, the *President's Award* from IPC at the 1990 Annual Conference, and recognition from several prominent professional organizations and technical societies. He has served as chair of the Product Design Sub-committee for Producibility Metrics sponsored by the United States Navy, and has served as the technical co-chair of the SPC standards committee for the Interconnecting and Packaging Electronics Circuits Institute (IPC). He has published in such journals as *IEEE*, *Micro and Circuit World*, and the *Journal of Circuit Technology*. He has written numerous technical papers and instructional manuals related to statistical engineering methods as well as a reference book on the application of experiment design, inferential statistics, and statistical process control procedures. Dr. Harry is a contributing author to a case-study textbook published by Marcel Decker and a textbook used by the Mathematics Department, U.S. Air Force Academy. He may be reached for consultation at (708) 538-2252, or by writing the Six Sigma Research Institute, 3701 East Algonquin Road, Suite 225, Rolling Meadows, Illinois 60008.

Jack Prins is a principal research scientist and manager of Research and Development at the Six Sigma Research Institute. He received his B.S. degree in electrical engineering from the New Jersey Institute of Technology, his M.S. degree in mathematical statistics from Rutgers University, and his Ph.D. in operations research from the New York University. Before joining the Motorola team, he was a statistical consultant to SEMATECH. Prior to this, he held various managerial and senior technical positions at IBM from 1966 to 1989. Dr. Prins has published several PC-based software packages in statistics, process control, and stockmarket analysis, as well as time-series analysis. In addition, he directed computer centers at the State University of New York and Vassar College, and served as an instructor for various courses in programming, mathematical statistics, numerical analysis, and time-series analysis. He has published in several major journals and has lectured extensively worldwide. During his long stay with IBM, he was credited with the kick-off of the entire IBM Information Center Marketing Program, as well as being the first developer of Statistical Software Packaging for the Host System. He commands several computer programming languages and speaks three foreign languages. Dr. Prins has professionally mentored many large corporations in the translation of statistical theory into practical application. He spent a sabbatical year at the Weizmann Institute in Israel, working on cancer research and implementation of VM/CMS in the Computer Center. Dr. Prins has been recognized on a number of occasions, culminating in IBM's Outstanding Contribution Award. He may be reached for consultation at (708) 538-2588, or by writing the Six Sigma Research Institute, 3701 East Algonquin Road, Suite 225, Rolling Meadows, Illinois 60008.

CHAPTER 1

One-Way Analysis of Variance

1 **DESCRIPTION**

   One-way analysis of variance, or *one-way ANOVA* as it is often termed, is a statistical method used to test the relationship between a given dependent variable and a single independent variable that is classified into two or more groups. Specifically, this procedure will ascertain whether or not the response means associated with the groups are drawn from the same population. The analysis involves arithmetically decomposing the total observed variation into two unique components. One component represents the response variation strictly attributable to the independent variable, and the other reports on *residual variation,* or *background noise* as it is sometimes called. With these estimates of variability, the one-way procedure evaluates the probability of equal component variances. If the probability exceeds a given threshold value, the alternate hypothesis of statistically significant difference (in component variances) would be accepted. Under this condition, it would be concluded that the observed variation in group means did not result from chance sampling variations.[1]

---

[1] The rationale for employing variances to test for significant differences in means is often a source of confusion. For a comprehensive discussion, the reader is directed to Box et al. (1978).

2   LIST OF EQUATIONS

Balanced One-way ANOVA (where $n_1 = n_2 = \ldots = n_g$)

$$\sum_{j=1}^{g} \sum_{i=1}^{n_j} \left(X_{ij} - \overline{\overline{X}} \right)^2 = n_j \sum_{j=1}^{g} \left(\overline{X}_j - \overline{\overline{X}} \right)^2$$
$$+ \sum_{j=1}^{g} \sum_{i=1}^{n_j} \left(X_{ij} - \overline{X}_j \right)^2, \qquad (1.1)$$

where

X_{ij} = the ith variate associated with the jth level or group,

g = the number of levels,

$\overline{\overline{X}}$ = the grand mean,

n_j = the total number of measurements within the jth level, and

$\overline{X}_j$ = the sampling mean of the jth level.

Unbalanced One-way ANOVA (for unequal n)

$$\sum_{j=1}^{g} \sum_{i=1}^{n_j} \left(X_{ij} - \overline{\overline{X}} \right)^2 = \sum_{j=1}^{g} n_j \left(\overline{X}_j - \overline{\overline{X}} \right)^2$$
$$+ \sum_{j=1}^{g} \sum_{i=1}^{n_j} \left(X_{ij} - \overline{X}_j \right)^2. \qquad (1.2)$$

3   VARIABLES

- The dependent variable is often called the *response*. It must be measured on an interval or ratio scale; e.g., pounds, inches, volts, temperature, etc.

- The independent variable is commonly known as a *factor*. In all cases, a factor must be treated as a categorical variable.

- If the factor is continuous by nature (i.e., if it is an interval or ratio), it must be classified into subgroups or *levels*. For example, if we were to consider two conditions of line pressure in a certain pipe, it would be said that this factor is represented by two levels.

- If the factor is discrete (i.e., nominal), further classification may not be required as the subgroups would exist naturally. For example, the variable called "coin" would be represented by two natural levels, i.e., heads and tails.

- If the factor levels are randomly selected from a larger set of levels, the resultant analysis accounts for *random effects* of the factor. In this case, conclusions could be drawn about those levels not considered in the analysis.

- If the factor levels are specifically assigned, subsequent analysis of the response would report on the *fixed effects* of the levels. Under this circumstance, conclusions are constrained to only those levels present in the analysis.

4   UNDERLYING ASSUMPTIONS

- Population variances of the response are equal across all levels of the given factor. This condition is referred to as *homogeneity of variance*. In many application scenarios, such an assumption is considered reasonable. If in doubt, there exist a variety of statistical methods that may assess equality of variances.[2]

---

[2] In practice, this assumption can be mildly violated with minimal effects. As a rule of thumb, the largest level variance should be no more than twice the size of the smallest level variance. If this holds, the Type I error will not change appreciably; e.g., a probability of 0.05 may degrade to 0.06 or, in extreme cases, to 0.07. To avoid gross changes in the Type I error probability, the effect of nonequal variances (heteroscedasticity) can be offset by an increase in sample size, the selection of a more conservative alpha criterion, and/or the equalization of observations across all factor levels (when applicable).

- Response means are independently and normally distributed. This is often considered to be a reasonable assumption by virtue of the Central Limit Theorem. It should be underscored that this particular assumption applies only to the distribution of sampling averages. The distribution of individual measurements does not have to be normally distributed, providing the distribution displays unimodality and is not markedly skewed.[3]

- If a model has been postulated, the errors are independently and normally distributed with mean of zero and a constant but finite variance.

- For each level of the factor, the response sample is representative of its corresponding population. If a random selection process is employed to draw the sample, this assumption is considered reasonable. Recognize that, as the given sample is diminished in size, the likelihood of representativeness also diminishes.

- Error resulting from measurement apparatus is negligible. If in doubt, a separate study should be undertaken to establish the extent and sources of such error and, if necessary, to reduce the error to an acceptable level.

5   APPLICATION CONSIDERATIONS

- The tool is quite robust to moderate violations of its underlying assumptions. As a result, the tool can be applied to a wide array of problems in which limited information is

---

[3] Recognize that mild skewedness can have a small effect on the Type I error probability. In the event of marked skewedness and inequality of variance, its effect on Type I error probability will most likely be strong. However, it must be remembered that the effects of skewedness can be significantly diminished via a mathematical transformation of the raw data. In addition, the likelihood of a gross change in the Type I error probability can be minimized by an increase in sample size, the selection of a more conservative alpha criterion, and/or the equalization of observations across all factor levels (when applicable).

available concerning the population(s) of interest. In most situations, the resulting analysis will provide sound information for decision making.

- Results can be easily reduced to common forms of data display; e.g., the bar graph and pie chart. As a consequence, the results can be easily communicated to a wide audience.

- Computational aspects of the tool are reasonably simple and straightforward.

- The tool is widely available on commercial software packages. In fact, some packages specialize in the general family ANOVA procedures.

- The tool can sometimes be used when the response data is discrete. However, in this circumstance, the researcher is well advised to seek the council of an experienced practitioner.

- The tool can be used in conjunction with a variety of statistically designed experiments.

6   APPLICATION SCENARIOS

The data referenced within this section and subsequently employed in the application roadmap is presented in Figure 1.1. The reader should be aware that the data are considered to

Level 1	Level 2	Level 3
10.0	18.0	21.0
12.0	15.0	18.0
13.0	14.0	15.0
14.0	17.0	12.0
19.0	15.0	16.0

FIGURE 1.1.   Coded Example Data Pertaining to the Application Scenarios

be in coded form. This perspective must be assumed so that the data retain application continuity across all of the scenarios. Given this, the reader should not try to relate the listed numbers to specific units of measure. For the sake of discussion, let it also be assumed that the given test conditions and sample sizes are appropriate.

Electrical Engineering: A certain electrical engineer was asked to select the most appropriate die size for a certain integrated circuit application. In this instance, the design engineer was concerned with IOH output current. After a close evaluation of the spec sheets, it was determined that there were only three different die sizes from which to select. It was generally believed that the three die sizes would perform in about the same manner. To confirm this, an appropriate test configuration was established. The response was generated by bench modeling the application. For each die size, the test was repeated five times. Subsequent to the test, the data were to be analyzed by the use of one-way analysis of variance. The resulting data are displayed in Figure 1.1.

Comment: The dependent variable is given as IOH output current. The independent variable is specified as die size. There are three different levels associated with the independent variable; i.e., the three different die sizes.

Mechanical Engineering: Mechnautics is a major manufacturer of electronic microcontrollers. Approximately four months ago they decided to bid on a contract involving the design and production of a controller much smaller than any they had previously manufactured. The performance requirements of this particular design were such that the controller could not be exposed to a temperature above 275°F. Because of its normal operating environment, the engineering team determined that the device would have to be artificially cooled. In this instance, the cooling options narrowed down to three; namely, the use of liquid, forced air, or a hybrid composite heat sink. To establish the feasibility of these cooling methods, an appropriate laboratory experiment was devised using 15 prototype devices, of which 5 were randomly assigned to each of the three treatment conditions. Based on the experiment design, it was decided that one-way analysis of variance should be employed to analyze the test data. The data are presented in Figure 1.1.

Comment: The dependent variable is given as the peak operating temperature of the device, under the prescribed experimental conditions. The independent variable is specified as cooling method. There are three different levels associated with the independent variable; i.e., the three types of cooling methods.

Process Engineering: Recently, XYZ Inc., a major manufacturer of printed circuit boards, received an alarming report related to a high field failure rate of its premier product, circuit board ABC. After a thorough analysis, it was concluded that the primary reason for failure was corrosion. After careful consideration of the evidence, it was hypothesized that the problem was related to the type of solder paste being used during production. Given this, a process engineer was tasked to evaluate the three types of solder pastes commonly used during production: RA, RMA, and SA. To make this evaluation, the engineer constructed a valid accelerated life test using 15 randomly selected printed wiring boards, each of which was randomly assigned to the three treatment conditions. It was determined that one-way analysis of variance should be used to analyze the resultant data. The data are located in Figure 1.1.

Comment: The dependent variable is specified as corrosion. The independent variable is defined as solder paste. RA, RMA, and SA are the three levels associated with the independent variable.

Software Engineering: Three major software houses have approached Bitpickers Inc. to obtain an exclusive licensing agreement for their respective C compilers. After discussions with each of the respective marketing departments, it was discovered that all three compilers were promoted as being the fastest in terms of processing speed. In order to investigate these claims, Bitpickers Inc. decided to conduct a trial. In this case, the trial was to be conducted using five randomly selected applications in conjunction with a common database. After consulting the local statistician, it was determined that one-way analysis of variance should be employed. The resulting test data are displayed in Figure 1.1.

Comment: The dependent variable is given as post-compiled processing speed. The independent variable is specified as compiler brand. There are three different levels associated with the independent variable, i.e., the three compiler versions.

Manufacturing: By nature of the business contracts, a certain manufacturing organization utilized three different workmanship standards relating to solder-joint quality. Essentially, they were using a typical commercial standard, Level I standard, and MIL-STD 2000. In the interests of standardization, the plant manager asked the manufacturing manager to investigate the differences between the three workmanship standards in terms of reportable defects. After some planning and consideration of the necessary controls, it was determined that the three standards should be evaluated using five randomly selected inspectors. It was also determined that one-way analysis of variance should be used to analyze the data. The resulting defect data are located in Figure 1.1.

Comment: The dependent variable is given as the number of reportable defects. The independent variable is specified as workmanship standard. There are three different levels associated with the independent variable, i.e., the three workmanship standards.

Administration: The corporate purchasing department of IJK, Inc. has recently implemented a new type of form (for internal use) related to the purchasing process. During the month following implementation of the new form, it was noted that the overall PO cycle time increased at one of the three production facilities. It was anticipated that the new forms would actually decrease overall cycle time. Needless to say, the increase greatly troubled the director of purchasing, especially since it was her idea to begin with. Interestingly, one of the managers on the director's staff pointed out that the increase in cycle time could be due to chance fluctuation rather than the new form. To test this construct, the purchasing director randomly selected five POs from each of the three facilities and recorded their respective cycle times. It was determined that the data should be analyzed using one-way analysis of variance. The resulting data are displayed in Figure 1.1.

Comment: The dependent variable is given as the overall purchasing cycle time. The independent variable is specified as facility location. There are three different levels associated with the independent variable, i.e., the three different facilities.

Facilities: The plant engineering department of a major corporation was concerned about the ability of its water purification process to meet a new internal standard related to parts-per-million contaminates. The plant engineer assigned to the problem quickly noted that the old process could be satisfactorily modified with the addition of a new type of filter. In this case, there were three manufacturers of the required filter. To establish which type of filter did the best job, an appropriate test was conducted. In essence, it involved the use of five randomly selected filters from each of the three suppliers. It was determined that the data should be analyzed using one-way analysis of variance. The resulting data are displayed in Figure 1.1.

Comment: The dependent variable is given as parts-per-million contaminates. The independent variable is specified as filter supplier. There are three different levels associated with the independent variable, i.e., the three different filter suppliers.

Finance: The corporate finance department of Deepockets Inc. regularly records and monitors time-card errors. During a recent meeting, the director of finance was asked if there were any appreciable differences in errors, pertaining to three different coding blocks on the time card, over a five-week period. The director was asked to report his findings by the end of the business day and, as a consequence, was greatly pushed for time. To do this, the director randomly sampled the names of 120 employees, duly noting the number of errors recorded for each of the three coding blocks. This was done for each of the five weeks under consideration using the same 120 employee names. It was determined that the data should be analyzed using one-way analysis of variance. The resulting data are displayed in Figure 1.1.

Comment: The dependent variable is given as the number of time-card errors. The independent variable is specified as time-card block. There are three different levels associated with the independent variable, i.e., the three different time-card blocks.

Personnel: The training organization of a major corporation regularly solicits participant feedback on its offerings. This is accomplished by using a standardized rating form. The form provides for the evaluation of the instructor and related course material across several different rating categories. To complete the form, the course participant indicates the extent (on a 1–5 scale) to which he or she agrees with the listed statements. Once completed, the forms are collected and subsequently filed with no regard to course title or instructor name. At a given point in time, the training manager was asked to establish whether or not there was any significant difference between the average rating of three different instructors. To do this, the manager opened the file cabinet drawer and arbitrarily selected five evaluation forms for each of the three instructors. It was determined that the data should be analyzed using one-way analysis of variance. The resulting data are displayed in Figure 1.1.

Comment: The dependent variable is given as the average rating score. The independent variable is specified as instructor. There are three different levels associated with the independent variable, i.e., the three different instructor names.

Sales: Company QRS Inc. manufactures and markets a certain type of two-way radio. The radio is only sold to city governments. In the recent past, several of QRS's customers stated that the addition of a new type of antenna would significantly enhance the product. Obviously, this had the potential to increase sales. In fact, the marketing department believed that such an addition would uniformly increase sales across all U.S. cities. To test this hypothesis, the vice president of marketing arbitrarily selected three major cities. Interestingly, these three cities accounted for 68 percent of the sales volume. In this instance, five customers were randomly selected within each of the three cities. In turn, the customers were polled using an interview-based survey instrument. It was determined that the resulting survey data should be analyzed using one-way analysis of variance. The data are displayed in Figure 1.1.

Comment: The dependent variable is given as the polling score. The independent variable is σ specified as city. There are three different levels associated with the independent variable, i.e., the three different cities.

7 APPLICATION ROADMAP

Step 1. Compute the sum for each level of the factor:

$$L_1 = \sum_{i=1}^{n_1=5} X_{i1} = 10 + 12 + 13 + 14 + 19 = 68,$$

$$L_2 = \sum_{i=1}^{n_2=5} X_{i2} = 18 + 15 + 14 + 17 + 15 = 79,$$

$$L_3 = \sum_{i=1}^{n_5} X_{i3} = 21 + 18 + 15 + 12 + 16 = 82.$$

Step 2. Compute the grand sum of all observations:

$$T = \sum_{j=1}^{g=3} \sum_{i=1}^{n_j=5} X_{ij} = 68 + 79 + 82 = 229.$$

Step 3. Compute the arithmetic mean of each level:

$$\overline{X}_1 = \frac{\sum_{i=1}^{n_1} X_{i1}}{n_1} = \frac{68}{5} = 13.6,$$

$$\overline{X}_2 = \frac{\sum_{i=1}^{n_2} X_{i2}}{n_2} = \frac{79}{5} = 15.8,$$

$$\overline{X}_3 = \frac{\sum_{i=1}^{n_3} X_{i3}}{n_3} = \frac{82}{5} = 16.4.$$

Step 4. Compute the grand mean:

$$\overline{\overline{X}} = \frac{T}{ng} = \frac{229}{5 \times 3} = 15.27.$$

Step 5. Compute the sum of all squared observations:

$$Q = \sum_{j=1}^{g} \sum_{i=1}^{n_j} X_{ij}^2 = 10^2 + 12^2 + \cdots + 16^2 = 3619.$$

Step 6. Compute the correction factor:

$$C = \frac{T^2}{ng} = \frac{229^2}{15} = 3496.07.$$

Step 7. Compute the total sum of squares:

$$SS_T = Q - C = 3619 - 3496.07 = 122.93.$$

Step 8. Compute the between sum of squares:

$$SS_B = \frac{\sum_{i=1}^{g=3} L_i^2}{n} - C = \frac{(68)^2 + (79)^2 + (82)^2}{5} - 3496.07 = 21.73.$$

Step 9. Compute the within sum of squares:

$$SS_W = SS_T - SS_B = 122.93 - 21.73 = 101.20.$$

Step 10. Construct the analysis of variance (ANOVA) table:

TABLE 1.1 *Generalized ANOVA Table*

Source of Variation	Sum of Squares (SS)	Degrees of Freedom (df)	Mean Square (MS)	MS Ratio (F_{calc})	Critical MSR (F_{crit})
(1) Between	(4) SS_B	(7) $g - 1$	(10) SS_B/df_B	(12) MS_B/MS_W	(13) F_{crit}
(2) Within	(5) SS_W	(8) $g(n - 1)$	(11) SS_W/df_W		
(3) Total	(6) SS_T	(9) $ng - 1$			

Interpretation: The critical F-value (F_{crit}) can be conveniently found in an F-table using the appropriate degrees of freedom and the specified probability of Type I error (α). If the calculated F-value (F_{calc}) is greater than or equal to F_{crit}, then accept the alternate hypothesis (H_a) with the specified degree of confidence $(1 - \alpha)$; otherwise, reject this hypothesis. Remember that the degree of confidence you have is always equal to one minus the risk you are willing to take in the commission of a Type I decision error. If the threshold value (F_{crit}) is not achieved, then simply reject H_a. As a general rule, H_o should not be accepted without knowing the probability of a Type II decision error (β). For the reader's convenience, the various terms within the ANOVA table are described as follows:

(1) = Factor source of variation.

(2) = Residual or error source of variation.

(3) = Total source of variation.

(4) = Factor sum of squares (reference Step 8). This is the sum of squares that can be assigned to the factor or "independent variable," as it is also called.

(5) = Residual or error sum of squares (reference Step 9). This is the sum of squares that can be assigned to the uncontrolled sources of variation, i.e., the variation that is due to the "background" variables.

(6) = Total sum of squares (reference Step 7). This is the sum of squares that can be assigned to the factor and background variables. The between-group and within-group sum of squares should total to the total sum of squares.

(7) = Between-level or between-group degrees of freedom.

(8) = Within-level or within-group degrees of freedom. Also called the residual degrees of freedom.

(9) = Total degrees of freedom. Note that the between- and within-degrees of freedom must sum to equal the total degrees of freedom.

(10) = Mean square of the factor.

(11) = Mean square of the residual.

(12) = Mean square ratio. The resulting number is also called the calculated F-value (F_{calc}).

(13) = Critical F-value (F_{crit}). The resulting number is also called the criterion or threshold value.

TABLE 1.2   *Example ANOVA Table*

Source of Variation	*Sum of Squares (SS)*	*Degrees of Freedom (df)*	*Mean Square (MS)*	*MS Ratio (F_{calc})*	*Critical MSR (F_{crit})*
Between	21.73	2	10.87	1.29	3.89
Within	101.20	12	8.43		
Total	122.93	14			

Interpretation: Since: F_{calc} (1.29) < F_{crit} (3.89), then reject H_a and conclude that the factor was not statistically significant.

Comment: The residual sum of squares is the variation that cannot be explained by variation in the independent variable. If we accept the fact that $Y = f(X_1 \ldots X_N)$ and subsequently allow X_1 to vary while holding $X_2 \ldots X_N$ constant, the effect of X_1 on the response variable (Y) can be assessed. However, it is most often not feasible or possible to hold so many of the other variables constant. As a result, we try just to control those variables that might exert the biggest unwanted effect. For the sake of argument, let us suppose that we were to control X_2, X_3, X_4, and X_5. This is done to remove their unique effects on the response variable. In other words, we would want to block their effect on the response. On the other hand, this means that variables $X_6 \ldots X_N$ would be uncontrolled during the experiment. This is one of the reasons why we randomize—to help ensure that any systematic effects that cannot be controlled are distributed among the factor levels in a relatively uniform manner. In short, this serves to ensure that sources of background noise and unwanted systematic influences are not predominant within any particular treatment. Let us further recognize that the total sum of squares includes the influence of all variables $X_1 \ldots X_N$ (with exception of the control variables). The factor sum of squares considers just the effect of the experimental variable (X_1). The difference between the total sum of squares and the experimental variable sum of squares has to be the influence exerted on the response variable by all other variables not controlled during the experiment (e.g., $X_6 \ldots X_N$). If the uncontrolled variables ($X_6 \ldots X_N$) fluctuate in a random fashion, their effect on the response variable would be relatively uniform across the experimental conditions of X_1. Because of this phenomenon, we observe within-treatment variation. The more variables that we control during the course of an experiment, the lower the within-treatment variation will be. As the within-treatment variation decreases, the residual sum of squares will likewise decrease. As a consequence, we are able to explain a larger portion of the total observed variation. When the explained variation increases, it becomes less likely that the experimental effect will be attributed to chance variation and, as a logical consequence, the likelihood of accepting the hypothesis of statistically significant difference will increase.

Step 11. Compute the relative effects:

$$R_B = 100\left(\frac{SS_B}{SS_T}\right) = 100\left(\frac{21.73}{122.93}\right) = 17.68\%.$$

Interpretation: The relative effect constitutes the proportion of explained variation. In other words, it is the percent of total variation accounted for by the factor. If the factor is not statistically significant, the relative effect should not be reported. Under this circumstance, it would be considered unexplained variation and, as a consequence, assigned to the error component. If the factor proves to be statistically significant, the relative effect should be reported as it provides the user with a desirable vehicle for communicating the analytical results to a statistically uninformed audience.

Step 12. Graph the relative effects:

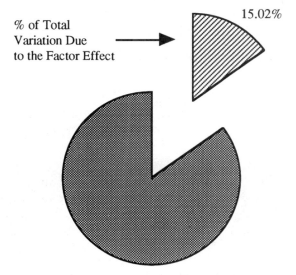

% of Total Variation Due to the Factor Effect

15.02%

FIGURE 1.2. Graph of the Relative Effects

Step 13. Construct a chart level for the means and standard error bars:

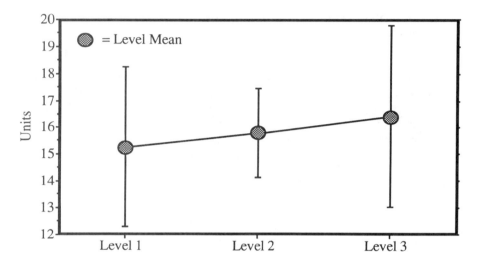

FIGURE 1.3.   One Standard Deviation Error Bars

Interpretation: Notice that the graph displays overlapping error bars. This condition is indicative of equal means, i.e., it illustrates the ANOVA results. Hence, it provides the user with an alternative vehicle to communicate the analytical results.

Step 14. Draw conclusions based on the analysis: For the given degrees of freedom, the factor was not statistically significant at the 95 percent level of confidence since the computed F-value (F_{calc}) did not exceed the critical F-value (F_{crit}). Therefore, it may be concluded that the observed differences between level means could have resulted from chance sampling variations rather than the differences in level settings. As a consequence, it may be said that the factor is not influential with respect to the response, i.e., the factor is among the "trivial many" versus the "vital few."

Comment: In general, if factor levels are set too close together, the factor effect can sometimes be unwittingly forced into a state of statistical insignificance. On the other hand, if the factor levels are set too far apart, the likelihood of accepting the hypothesis of statistically significant difference (H_a) will increase. To a large extent, statistical significance is highly dependent upon the level settings of the experimental variable. This is why it is so important to compute the proper sample size in advance of the study, using α, β, and $\delta\sigma$ as a guide for meshing practical difference with statistical significance. In addition, statistical significance is also (in many instances) highly dependent upon the number of factors that are controlled during the course of experimentation. If numerous variables are artificially controlled, the residual sum of squares will be lower, which, in turn, will force the resulting MSR to be higher. In short, the use of control variables can, to some extent, artificially inflate the statistical significance of an experimental variable as well as the relative effects (the percent of explained variation). This is only one of the reasons why some people say that "you can get statistics to say what ever you want them to say." As a consequence, the novice practitioner is counseled to seek expert guidance in those experiments or comparative situations where the consequences of decision error are intolerable.

REFERENCES AND BIBLIOGRAPHY

Box, G. E. P., W. G. Hunter, and J. S. Hunter. 1978. *Statistics for Experimenters*. New York: John Wiley and Sons.

Juran, J. M., F. M. Gryna, and R. S. Bingham. 1974. *Quality Control Handbook*, 3rd Edition. New York: McGraw-Hill.

CHAPTER 2

Two-Way Analysis of Variance

1 DESCRIPTION

The two-way analysis of variance can be used to analyze the simplest of factorial designs in which the experimenter is studying the main effects and potential interactive effects of two factors. In this experiment, a complete trial or replication is observed for all possible combinations of the levels of the factors being investigated. The two-way analysis of variance partitions the variability into a source of variability for the main effects, the interaction, and the error. The fixed-effects analysis of variance gives the user information that is applicable only to the levels of the factors that the experimenter has chosen.

2 LIST OF EQUATIONS

Sum of Squares Identity for a Balanced Two-Way Design[1]

$$\sum_{i=1}^{a} \sum_{j=1}^{b} \sum_{k=1}^{n} (y_{ijk} - \bar{y}...)^2$$

$$= bn \sum_{i=1}^{a} (\bar{y}i.. - \bar{y}...)^2 + an \sum_{j=1}^{b} (\bar{y}_{.j.} - \bar{y}...)^2$$

$$+ n \sum_{i=1}^{a} \sum_{j=1}^{b} (\bar{y}_{ij.} - \bar{y}_{i..} - \bar{y}_{.j.} - \bar{y}_{...})^2$$

$$+ \sum_{i=1}^{a} \sum_{j=1}^{b} \sum_{k=1}^{n} (y_{ijk} - \bar{y}_{ij.})^2,$$

where

y_{ijk} = the nth observation of the ith level of the treatment A and the jth level of treatment B,

$\bar{y}_{ij.}$ = the mean of the ijth cell,

$\bar{y}_{i..}$ = the mean of the ith level of treatment A,

$\bar{y}_{.j.}$ = the mean of the jth level of treatment B,

$\bar{y}_{...}$ = the grand mean of all abn observations,

a = the number of levels of treatment A,

b = the number of levels of treatment B, and

n = the number of replications.

[1] The equations given are for the case of the balanced design, where the number of observations is the same for each experimental combination. If an experimental observation is lost, broken, or destroyed, the case is an unbalanced case. There are two ways to handle the unbalanced case. The first is to treat the experiment as if there are missing data. This can be done if the number of observations missing is small. There exist some techniques for handling this situation, which are outlined in Montgomery (1991). If the experimenter has at least one observation per cell but an unequal number of observations, then the equations for the unbalanced analysis of variance should be used.

Computational Formulas for the ANOVA of the Balanced Case

$$SST = \sum_{i=1}^{a} \sum_{j=1}^{b} \sum_{k=1}^{n} y_{ijk}^2 - \frac{y_{...}^2}{abn},$$

$$SSA = \sum_{i=1}^{a} \frac{y_{i..}^2}{bn} - \frac{y_{...}^2}{abn},$$

$$SSB = \sum_{j=1}^{b} \frac{y_{.j.}^2}{an} - \frac{y_{...}^2}{abn},$$

$$SSAB = \sum_{i=1}^{a} \sum_{j=1}^{b} \frac{y_{ij.}^2}{n} - \frac{y_{...}^2}{abn} - SSA - SSB,$$

where

SST = the total sum of squares,
SSA = the sum of squares for treatment A,
SSB = the sum of squares for treatment B,
$SSAB$ = the sum of squares for the AB interaction,
y_{ijk} = the nth observation of the ith level of the treatment A and the jth level of treatment B,
$y_{ij.}$ = the mean of the ijth cell,
$y_{i..}$ = the mean of the ith level of treatment A,
$y_{.j.}$ = the mean of the jth level of treatment B,
$y_{...}$ = the grand mean of all abn observations,
a = the number of levels of treatment A,
b = the number of levels of treatment B, and
n = the number of replications.

3   VARIABLES

- The input or independent variable is called a *factor* or a *treatment*. The independent variable is controlled by the experimenter for fixed-effects models. The independent variable is a *categorical variable*.

- The dependent variable is called a *response variable*. The response variable is the measured or the observed value from the experiment.

- The independent variables may be discrete or continuous. When discrete, the factor is separated naturally into levels. For example, two different mold compounds have two natural levels or groups. Continuous variables (i.e., temperature, pressure, time, etc.) need to have levels established by the experimenter.

4 UNDERLYING ASSUMPTIONS

- The errors of the linear model associated with the two-way analysis of variance are normally and independently distributed with a mean of zero and a constant variance.

- The variance of the response is assumed to be constant for all levels of a given factor. The analysis of variance is robust to mild violations of equivalence of variance. As a rule of thumb, if the ratio of the highest variance to the lowest variance is less than 2, then there are negligible fluctuations in the analysis of variance. If there is a question about equality of variance, Bartlett's test will be useful. Details of this test can be found in Montgomery (1991).

- The observations of the ijth cell are a random sample of size n from a normally distributed population with mean μ_{ij} and variance σ^2.

- Input variables or factors must be independent of each other.

- The experiment is run in a random order. Randomization can be accomplished by using a random number table or other technique.

- The conclusions of a fixed effects model apply only to the levels of the factors included in the experiment. The conclusions cannot be extended to include any level not explicitly in the experiment.

5 APPLICATION CONSIDERATIONS

- Several analysis of variance software packages exist for personal computers. Analysis of variance routines can be found in statistical packages as well as design of experiments packages. The sheer number of computations goes up geometrically with the number of variables added to the analysis of variance. Without the use of a software package, it is very easy to make an arithmetic error.

- For single replicated designs (designs where there is only one observation per experimental combination) there are no degrees of freedom in which to compute an error sum of squares (SSE). These problems can be handled with Daniel's Normal Probability Plot. For further details refer to Montgomery (1991).

- If all of the factors are at only two levels, a high and a low level, then the manual computations can be greatly reduced by using Yates' Algorithm. For further details refer to any of the references. Yates also has an algorithm if all of the factors are at three levels.

- While equality of variance is an assumption in the analysis of variance, variances can and should be checked with residual plots. Residual plots of the factors, residual plots of the experiment over time, and residual plots of the actual observed values versus the predicted values should be checked for outliers as well as any patterns. These residual plots should be structureless. A normal probability plot of the residuals is also recommended to check the normality assumption about the residuals. Good computer programs will have the ability to generate these plots easily.

- Coding observations can simplify the calculations and improve the accuracy in the analysis of variance. Basic operations of addition, subtraction, multiplication, and division can be performed on each of the individual observations of an experiment without changing the resulting F-ratios of the analysis of variance. For example, if all of the observations of an experiment are of the form $0.xx$, then they can all be multiplied by 100 to remove the decimal point. This transformation will make the numbers easier to use in the computations and help reduce possible round-off errors. The F-ratios for the coded data will be the same as the F-ratios for the uncoded data. It is not generally necessary to code the data if a computer program is being used.

- Data transformations can be used to stabilize the observations of an experiment with respect to variance. Some common transformations, in order of strength, are the square root, the logarithmic, the reciprocal square root, and the reciprocal. Poisson or count data are best handled with a square root transformation. For more information on data transformations in the analysis of variance, refer to Montgomery (1991).

6   APPLICATION SCENARIOS

The data in Figure 2.1 will be referred to in each of the following applications scenarios as well as the application roadmap. The data are supplied for the sake of illustration.

	A1	*A2*	*A3*
B1	6.7	2.7	1.4
	6.1	4.8	2.4
B2	5.4	2.1	3.8
	5.9	1.9	3.4

FIGURE 2.1.   Coded Data for Application Scenarios

Assume that the data are in coded form; do not try to associate specific units or meaning to numbers. Also assume for the sake of illustration that the sampling size is appropriate and was taken in a random fashion.

Electrical Engineering: An electrical engineer at Sili Semiconductor, Inc. wanted to reduce the Vcb(f) in a bipolar power transistor by comparing three different back metal schemes and two different substrates. The three back metal schemes are TiNiAg, AlTiNiAg, and CrNiAg. The substrates are antimony and arsenic. The experimental design that the engineer chose to use was a replicated two-by-three factorial experiment. Each cell in the experiment was the average of 27 readings, (three wafers per cell with nine readings per wafer). The engineer measured the Vcb(f), calculated the averages, and performed a two-way analysis of variance (ANOVA). The resulting coded data are displayed in Figure 2.1.

Mechanical Engineering: A mechanical engineer at IC, Inc. has been instructed by his supervisor to improve the hermeticity of the plastic packages that the company manufactures for their integrated circuit product families. After investigating the competitions' products, the engineer decided to run an experiment that included three different mold compounds and two heat-sink plating materials. He chose to run 24 units per cell and used the average of the die-penetrant measurements to represent the cell. The engineer chose to replicate the experiment once so as to establish an estimate for error in his experiment. He realized that he had a two-by-three factorial experiment and that a two-way ANOVA would be required to analyze his experiment. After he assembled the units, he performed a die-penetrant test and measured the distance that the die penetrated into the package. The resulting coded data are displayed in Figure 2.1.

Process Engineering: A process engineer for a semiconductor company is interested in utilizing a plasma etch system as an oxide etch. The etcher is capable of using three gases, CF4, SF6, and NF3. The plasma etch system has a low and high setting for the power. The engineer designed a replicated two-by-three factorial experiment to evaluate the gases as well as the power level. The response the engineer was interested in was uniformity across the wafer. A two-way analysis of variance was used to analyze the uniformity results. The resulting data are displayed in Figure 2.1.

Software Engineering: Because of customer complaints, a software engineer at Wiz Bytes, Inc., was instructed to reduce the CPU time required to run an analytical software package called Number Cruncher. The primary tool that this package offers is solutions to complicated simultaneous equations. The engineer chose to investigate three different languages, Fortran, C, and Pascal. She also chose to try two different methods for solving the equations, Gauss elimination, and the fundamental theorem of linear algebra. Because the software is designed for mainframe computers with multiple users, she decided to replicate the experiment once to allow for error in the system. She derived a problem that required the solution of 100 equations for 100 variables. After running the experiment, she performed a two-way ANOVA to analyze the data. The resulting data are displayed in Figure 2.1.

Manufacturing: Due to the results of a government audit, the general manager of Cream Works dairy products facility was determined to standardize his operations. This included the clean-room garb that the employees wear when working inside the homogenization area. Rather than arbitrarily picking which gowns and hair nets the operators would use, he decided to let a manufacturing engineer perform a study to determine which garments were the best in preventing particles and other contamination from getting in the product. There were three types of gowns and two hair nets, blue and white. The engineer decided to use himself as the test vehicle for his experiment. He would put on the garments and stand next to a particle counter for 15 minutes. He did this for every combination of gowns and hair nets. So as to have an estimate of the error, he chose to replicate his experiment once. The experiment that he designed was a two-by-three factorial experiment and he performed a two-way analysis of variance to analyze the data. The resulting data are displayed in Figure 2.1.

Administration: The secretary of the engineering office at Jerco Refrigeration, Inc., was told by many that the memos she typed were not being read because they were too difficult to read. She printed her work on a very good laser writer and did not believe that the printer was the problem. Instead, she decided to evaluate the font that she used when typing the memos. She was also being told by the operations manager for engineering to make a concerted effort to conserve paper by using one page memos whenever possible. She decided to run a replicated three-by-two factorial design so as to address both of these issues. The secretary chose three different fonts and two different font sizes for her experiment. She randomly selected a group of engineers to be on a memo examination board to grade her memos on a scale of 1 to 100 where the lower the score, the better. She typed the memo 12 different times, using a different combination of fonts and sizes for each memo as well as replicating each. She mixed up the memos and distributed them to the board such that every engineer read every memo and scored the memos appropriately. After the examinations of the memos, the secretary averaged the scores for each memo and divided the average by the number of the characters per page of the memo. She analyzed her results by using a two-way ANOVA. The resulting data are displayed in Figure 2.1.

Facilities: Acme Optics, Inc., has been enjoying the lion's share of the high-grade laser-quality optics market for several years. However, as in many industries, their competition has recently been selling better-quality products at lower costs. Upon reviewing the Pareto of defects, it was obvious that the major quality problem at Acme was particulate contamination. The facility engineering department was immediately made aware of this issue, and hence put their best engineer on the task. The manufacturing areas have always had sufficient laminar flow hoods with the filters changed on a regular basis. However, the kind of filters that they were using had been around for several years and the engineer decided to evaluate some of the other filters that were available. He chose to compare two other filters along with the one that they had been using. The laminar flow hoods have low and high settings for air flow. The engineer decided to run a replicated two-by-three factorial design by changing the air-flow settings and the filters. He measured the particle count for 15 minutes under the hood for each set of conditions. The engineer analyzed the experiment by using a two-way analysis of variance. The resulting data are displayed in Figure 2.1.

Finance: The accounting office of Tite and Waud Investments, Inc., was interested in increasing the efficiency of the pencils that they use in the office. They decided to run a two-by-three factorial experiment to evaluate three different types of lead as well as two different pencil sharpeners. The experiment would be replicated only once for the estimate of error. Mr. Tite did not want to spend any more money on the experiment than necessary. Using a constant pressure, the sharpened pencil would be used to draw a straight line until the width of the line was greater than a sixteenth of an inch wide. At this point, the pencil would be sharpened again by the same sharpener that was used on this pencil earlier, and would begin drawing another line. The pencil is expired when it reaches two-and-one-half inches in length. The sum of the lengths of the lines for each pencil/sharpener combination is recorded. A two-way analysis was used to analyze the data. The resulting data are displayed in Figure 2.1.

Personnel: There was an extremely serious problem for the personnel office of GPL Electronics, Inc., that required the undivided attention of the entire department for three months. The problem was that none of the 15,000 employees were reading the employee newsletter. To try to bring about a new interest in the newsletter, the personnel department chose to evaluate three different colors for paper, red, green, and blue, on which to print the newsletter, and to evaluate black and white inks. The six cells of the experiment were all replicated once for a total of twelve experimental runs. The run order was determined randomly. Each run was to be performed once a week per the regular circulation of the letter. Each week 15,000 newsletters were published and put out into the newsletter stands at the entrances to the facility. Personnel recorded the number of newsletters remaining after the second day of publication. A two-way analysis of variance was used to analyze the data. The resulting data are displayed in Figure 2.1.

Sales: A marketer for White Laundry Cosmetics was asked by the marketing manager to redesign their cosmetic case, which is free with the purchase of 28 dollars or more. In the cosmetics industry, the freebies serve a more important role in the sale of products than the product itself. Rather than arbitrarily picking a new design, the marketer decided to let the customers pick it for her. She decided to run a two-by-three factorial design to compare a zipper closure with that of a snap as well as three different patterns, floral, dots, and plaid. The experiment would take place at a very popular department store on a Saturday and Sunday. The customers would be given a choice of the six different cases before they made their purchase. At the close of each day, the marketer recorded the number of cosmetic cases in each group that were given away. She then performed a two-way ANOVA on her data. The resulting data are displayed in Figure 2.1.

7   APPLICATION ROADMAP

Step 1. State the risk level:

$$\alpha = 0.05, \text{ or equivalently } \alpha = 5\%.$$

Step 2. Computations. Compute the totals for each treatment, each cell, and for all observations.

Step 2.1. Compute the totals for each treatment of factor $A(y_{i..})$ and record this number to the right of the corresponding treatment row:

$$y_{1..} = 6.7 + 6.1 + 5.4 + 5.9 = 24.1.$$

$$y_{2..} = 2.7 + 4.8 + 2.1 + 1.9 = 11.5.$$

$$y_{3..} = 1.4 + 2.4 + 3.8 + 3.4 = 11.0.$$

Step 2.2. Compute the totals for each treatment of factor $B(y_{.j.})$ and record this number at the bottom of the corresponding treatment column.

$$y_{.1.} = 6.7 + 6.1 + 2.7 + 4.8 + 1.4 + 2.4 = 24.1.$$

$$y_{.2.} = 5.4 + 5.9 + 2.1 + 1.9 + 3.8 + 3.4 = 22.5.$$

Step 2.3. Compute the total for each cell $(y_{ij.})$ and record this value in a circle next to the corresponding cell as shown in Figure 2.2.

$$y_{11.} = 6.7 + 6.1 = 12.8.$$

$$y_{12.} = 5.4 + 5.9 = 11.3.$$

$$y_{21.} = 2.7 + 4.8 = 7.5.$$

$$y_{22.} = 2.1 + 1.9 = 4.0.$$

$$y_{31.} = 1.4 + 2.4 = 3.8.$$

$$y_{32.} = 3.8 + 3.4 = 7.2.$$

Step 2.4. Compute the grand total $(y...)$ by adding the totals of treatment A or by adding the totals of treatment B. It is a good arithmetic check to calculate the grand totals both ways. Record at the bottom right of the data:

Using the totals of A: $y... = 24.1 + 11.5$
$$+ 11.0 = 46.6$$
Using the totals of B: $y... = 24.1 + 22.5$
$$= 46.6$$

	A1	*A2*	*A3*	$y_{j.}$
B1	6.7 (12.8) 6.1	2.7 (7.5) 4.8	1.4 (3.8) 2.4	24.1
B2	5.4 (11.3) 5.9	2.1 (4.0) 1.9	3.8 (7.2) 3.4	22.5
$y_{i..}$	24.1	11.5	11.0	$y...$ 46.6

FIGURE 2.2.   Coded Data with Factor and Cell Totals

Step 3. Construct the ANOVA Table. Replace the entries in Table 2.1 with the computed values.

Step 4. Compute the degrees of freedom for A, B, AB, error and total.

Step 4.1. Compute the degrees of freedom for treatment A and record in the ANOVA table.

$$\text{degrees of freedom } (A) = \text{number of } A \text{ treatments} - 1 = a - 1.$$
$$\text{degrees of freedom } (A) = 3 - 1 = 2.$$

Step 4.2. Compute the degrees of freedom for treatment B and record in the ANOVA table.

$$\text{degrees of freedom } (B) = \text{number of } B \text{ treatments} - 1 = b - 1.$$
$$\text{degrees of freedom } (B) = 2 - 1 = 1.$$

Step 4.3. Compute the degrees of freedom for the AB interaction and record in the ANOVA table.

$$\text{degrees of freedom } (AB) = (a - 1)(b - 1).$$
$$\text{degrees of freedom } (AB) = (3 - 1)(2 - 1)$$
$$= 2 \times 1 = 2.$$

Step 4.4. Compute the degrees of freedom for error and record in the ANOVA table.

$$\text{degrees of freedom (error)} = ab \text{ (number of replicates} - 1)$$
$$= ab(n - 1).$$
$$\text{degrees of freedom (error)} = (3 \times 2)(2 - 1)$$
$$= 6 \times 1 = 6.$$

Step 4.5. Compute the degrees of freedom for the total and record in the ANOVA table. Add the degrees of freedom for the treatments, blocks, and error. This

TABLE 2.1 *General ANOVA Table*

Source of Variation	Degrees of Freedom	Sum of Squares	Mean Square	F	F Critical
A Treatments	$a - 1$	SSA	$MSA = \dfrac{SSA}{a - 1}$	$\dfrac{MSA}{MSE}$	$F_{\alpha,(a-1),ab(n-1)}$
B Treatments	$b - 1$	SSB	$MSB = \dfrac{SSB}{b - 1}$	$\dfrac{MSB}{MSE}$	$F_{\alpha,(b-1),ab(n-1)}$
AB Interaction	$(a - 1)(b - 1)$	SSAB	$MSAB = \dfrac{SSAB}{(a - 1)(b - 1)}$	$\dfrac{MSAB}{MSE}$	$F_{\alpha,(a-1)(b-1),ab(n-1)}$
Error	$ab(n - 1)$	SSE	$MSE = \dfrac{SSE}{ab(n - 1)}$		
Total	$abn - 1$	SST			

sum should equal the degrees of freedom for the total.

$$\text{degrees of freedom (total)} = \text{total number of}$$
$$\text{observations} - 1$$
$$= abn - 1.$$
$$\text{degrees of freedom (total)} = (3 \times 2 \times 2) - 1$$
$$= 12 - 1 = 11.$$

Step 5. Compute the sum of squares.

Step 5.1. Compute the correction factor:

$$C = \frac{y_{...}^2}{abn} = \frac{46.6^2}{12} = 180.9633.$$

Step 5.2. Compute the sum of squares total (SST) and record in the ANOVA table.

$$SST = \sum_{i=1}^{a} \sum_{j=1}^{b} \sum_{k=1}^{n} y_{ijk}^2 - C,$$

$$SST = (6.7^2 + 6.1^2 + 2.7^2 + 4.8^2 + 1.4^2$$
$$+ 2.4^2 + 5.4^2 + 5.9^2 + 2.1^2 + 1.9^2$$
$$+ 3.8^2 + 3.4^2) - C,$$
$$SST = 218.14 - 180.9633 = 37.177.$$

Step 5.3. Compute the sum of squares for the A treatments (SSA) and record in the ANOVA table:

$$SSA = \sum_{i=1}^{a} \frac{y_{i..}^2}{bn} - C,$$

$$SSA = \frac{(24.1^2 + 11.5^2 + 11.0^2)}{4} - C,$$

$$SSA = \frac{834.06}{4} - 180.9633 = 208.515$$

$$- 180.9633 = 27.5517.$$

Step 5.4. Compute the sum of squares for the B treatments (SSB) and record in the ANOVA table:

$$SSB = \sum_{j=1}^{b} \frac{y_{.j.}^2}{an} - C,$$

$$SSA = \frac{(24.1^2 + 22.5^2)}{6} - C,$$

$$SSA = \frac{1087.06}{6} - 180.9633 = 181.1767$$

$$- 180.9633 = 0.2134.$$

Step 5.5. Compute the sum of squares of the interaction AB ($SSAB$) and record in the ANOVA table:

$$SSAB = \sum_{i=1}^{a} \sum_{j=1}^{b} \frac{y_{ij.}^2}{n} - SSA - SSB - C,$$

$$SSAB = \frac{(12.8^2 + 11.3^2 + 7.5^2 + 4.0^2 + 3.8^2 + 7.2^2)}{2}$$

$$- SSA - SSB - C,$$

$$SSAB = \frac{430.06}{2} - 27.5517 - 0.2134 - 180.9633,$$

$$SSAB = 215.03 - 27.5517 - 0.2134 - 180.9633 = 6.3016.$$

Step 5.6. Compute the sum of squares for the error (SSE) and record in the ANOVA table:

$$SSE = SST - SSA - SSB - SSAB$$
$$= 37.1767 - 27.5517 - 0.2134$$
$$- 6.3016 = 3.11.$$

Step 6. Compute the mean squares.

Step 6.1. Compute the mean square treatment A (MSA) and record in the ANOVA table.

$$MSA = \frac{SSA}{\text{degrees of freedom for treatment } A}$$

$$= \frac{27.5517}{2} = 13.7759.$$

Step 6.2. Compute the mean square of treatment B (MSB) and record in the ANOVA table.

$$MSB = \frac{SSB}{\text{degrees of freedom for treatment } B}$$

$$= \frac{0.2134}{1} = 0.2134.$$

Step 6.3. Compute the mean square of interaction AB ($MSAB$) and record in the ANOVA table.

$$MSAB = \frac{SSAB}{\text{degrees of freedom for interaction } AB}$$

$$= \frac{6.3016}{2} = 3.1508$$

Step 6.4. Compute the mean square error (MSE) and record in the ANOVA table.

$$MSE = \frac{SSE}{\text{degrees of freedom for error}}$$

$$= \frac{3.11}{6} = 0.5183.$$

Step 7. Compute the F-values.

Step 7.1. Compute the F-value for treatment A and record in the ANOVA table.

$$F(A) = \frac{MSA}{MSE} = \frac{13.7759}{0.5183} = 26.5790.$$

Step 7.2. Compute the F-value for treatment B and record in the ANOVA table.

$$F(B) = \frac{MSB}{MSE} = \frac{0.2134}{0.5183} = 0.4117.$$

Step 7.3. Compute the F-value for interaction AB and record in the ANOVA table. After this step, the ANOVA table should look like Table 2.2.

$$F(AB) = \frac{MSAB}{MSE} = \frac{3.1508}{0.5183} = 6.0791.$$

TABLE 2.2   *Complete ANOVA Table for Scenario Data*

Source of Variation	Degrees of Freedom	Sum of Squares	Mean Square	F
A	2	27.5517	13.7759	26.5790*
B	1	0.2134	0.2134	0.4117
AB	2	6.3016	3.1508	6.0791*
Error	6	3.1100	0.5183	
Total	11	37.1170		

Step 8. Determine the critical F-regions. F-tables can be found in all the references as well as most statistics books.

Step 8.1. Determine the critical F-region for the *A* treatments. Look up F_{α,ν_1,ν_2} in an F-table; where α is the risk factor, ν_1 is the number of degrees of freedom for the *A* treatments, and ν_2 is the number of degrees of freedom for the error.

$$F_{\alpha,\nu 1,\nu 2} = F_{0.05,2,6} = 5.14.$$

Step 8.2. Determine the critical F-region for the *B* treatments. Look up F_{α,ν_1,ν_2} in an F-table; where α is the risk factor, ν_1 is the number of degrees of freedom for the *B* treatments, and ν_2 is the number of degrees of freedom for the error.

$$F_{\alpha,\nu_1,\nu_2} = F_{0.05,1,6} = 5.99.$$

Step 8.3. Determine the critical F-region for the *AB* interaction. Look up F_{α,ν_1,ν_2} in an F-table; where α is the risk factor, ν_1 is the number of degrees of freedom for the *AB* interaction, and ν_2 is the number of degrees of freedom for the error.

$$F_{\alpha,\nu_1,\nu_2} = F_{0.05,2,6} = 5.14.$$

Step 9. Draw Conclusions: Compare the calculated F-value to the critical F-value for each of the main effects and interactions. The significant effects are

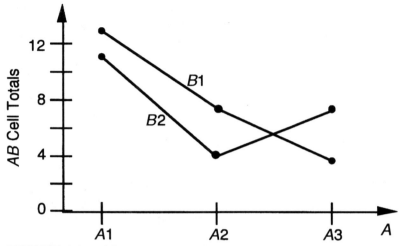

FIGURE 2.3.   *AB* Interaction Plot for Scenario Data

those where the calculated F-value is greater than the critical F-value. In this example,

$$F(A) = 26.5790 > F_{0.05,2,6} = 5.14, \text{ and}$$

$$F(AB) = 6.0791 > F_{0.05,2,6} = 5.14.$$

The F-values of the remaining effects are less than their corresponding critical F-values. The significant effects for this example are *A* and the *AB* inter-action. Since the *AB* interaction was found to be significant, it is useful to view this interaction graphically.

The graph of the *AB* interaction shows that at the low level of *B* (*B*1) the output is nearly linear. At the high level of *B* (*B*2) the response over the factor *A* is more parabolic. The experimenter should use the information in an interaction plot to advantage. For example, if the desired response is low, then the experimenter can choose to run the process at the middle level of *A* and the high level of *B*, or the high level of A and the low level of B. Cost, safety, and process sensitivity are some of the other factors that should be considered when choosing between the two alternatives.

The experimenter also should check model adequacy with residual plots.

REFERENCES AND BIBLIOGRAPHY

Hicks, C. R. 1964. *Fundamental Concepts in the Design of Experiments*. New York: Holt, Rinehart & Winston.

Hines, W. W., and D. C. Montgomery. 1991. *Probability and Statistics in Engineering and Management Science*, 3rd Edition. New York: John Wiley and Sons.

Miller, I., and J. E. Freund. 1985. *Probability and Statistics for Engineers*, 3rd Edition. Englewood Cliffs, New Jersey: Prentice-Hall.

Montgomery, D. C. 1991. *Design and Analysis of Experiments*, 3rd Edition. New York: John Wiley and Sons.

Walpole, R. E., and R. H. Myers. 1989. *Probability and Statistics for Engineers and Scientists*, 4th Edition. New York: MacMillan.

CHAPTER 3

N-Way Fixed Analysis

1 DESCRIPTION

The N-way analysis of variance can be used to analyze factorial designs in which the experimenter is studying the main effects and potential interactive effects of several factors. In this analysis, a complete trial or replication is observed for all possible combinations of the levels of the factors being investigated. The N-way analysis of variance partitions the variability into a source of variability for the main effects, the interactions, and the error. The fixed-effects analysis of variance gives the user information that is applicable only to the levels of the factors that the experimenter has chosen.

2 LIST OF EQUATIONS

Computational Formulas for the N-way Analysis of Variance:
Balanced Case[1]

$$SST = \sum_{i=1}^{a} \sum_{j=1}^{b} \sum_{k=1}^{c} \sum_{l=1}^{n} y_{ijkl}^2 - \frac{y_{....}^2}{abcn},$$

---

[1] The equations given are for the case of the balanced design where the number of observations is the same for each experimental combination. If an experimental observation is lost, broken, or destroyed, the case is an unbalanced case. There are two ways to handle the unbalanced case. The first is to treat the experiment as if there are missing data. This can be done if the number of observations missing is small. There exist some techniques for handling this situation, outlined in Montgomery (1991). If the experimenter has at least one observation per cell but an unequal number of observations, then the equations for the unbalanced analysis of variance should be used.

$$SSA = \sum_{i=1}^{a} \frac{y_{i...}^2}{bcn} - \frac{y_{....}^2}{abcn},$$

$$SSAB = \sum_{i=1}^{a} \sum_{j=1}^{b} \frac{y_{ij..}^2}{cn} - \frac{y_{....}^2}{abcn} - SSA - SSB,$$

$$SSB = \sum_{j=1}^{b} \frac{y_{.j..}^2}{acn} - \frac{y_{....}^2}{abcn},$$

$$SSAC = \sum_{i=1}^{a} \sum_{k=1}^{c} \frac{y_{i.k.}^2}{bn} - \frac{y_{....}^2}{abcn} - SSA - SSC,$$

$$SSC = \sum_{k=1}^{c} \frac{y_{..k.}^2}{abn} - \frac{y_{....}^2}{abcn},$$

$$SSBC = \sum_{j=1}^{b} \sum_{k=1}^{c} \frac{y_{.jk.}^2}{an} - \frac{y_{....}^2}{abcn} - SSB - SSC, \text{ and}$$

$$SSABC = \sum_{i=1}^{a} \sum_{j=1}^{b} \sum_{k=1}^{c} \frac{y_{ijk.}^2}{n} - \frac{y_{....}^2}{abcn} - SSA - SSB$$
$$- SSAB - SSC - SSAC - SSBC,$$

where

SST	= the total sum of squares,
SSA	= the sum of squares for treatment A,
SSB	= the sum of squares for treatment B,
$SSAB$	= the sum of squares for the AB interaction,
SSC	= the sum of squares for treatment C,
$SSAC$	= the sum of squares for the AC interaction,
$SSBC$	= the sum of squares for the BC interaction,
$SSABC$	= the sum of squares for the ABC interaction,
y_{ijkl}	= the nth observation of the ith level of the treatment A, the jth level of treatment B and the kth level of treatment C,
$y_{i...}$	= the total of the ith level of treatment A,
$y_{.j..}$	= the total of the jth level of treatment B,
$y_{..k.}$	= the total of the kth level of treatment C,
$y_{ij..}$	= the total of the ith level of the treatment A and the jth level of treatment B,

$y_{i.k.}$ = the total of the ith level of the treatment A and the kth level of treatment C,

$y_{.jk.}$ = the total of the jth level of the treatment B and the kth level of treatment C,

$y_{ijk.}$ = the total of the ijkth cell,

$y...$ = the grand total of all abn observations,

a = the number of levels of treatment A,

b = the number of levels of treatment B,

c = the number of levels of treatment C, and

n = the number of replications.

3   VARIABLES

■ The input or *independent variable* is called a *factor* or a *treatment*. The independent variable is controlled by the experimenter (for fixed-effects models). The independent variable is a categorical variable.

■ The *dependent variable* is called a *response variable*. The response variable is the measured or observed value from the experiment.

■ The independent variables may be discrete or continuous. When discrete, the factor is naturally separated into levels. For example, two different mold compounds have two natural levels or groups. Continuous variables (i.e., temperature, pressure, time, etc.) need to have levels established by the experimenter.

4   UNDERLYING ASSUMPTIONS

■ The errors of the linear model associated with the N-way analysis of variance are normally and independently distributed with a mean of zero and a constant variance.

- The variance of the response is assumed to be constant for all levels of a given factor. The analysis of variance is robust to mild violations of equivalence of variance. As a rule of thumb, if the ratio of the highest variance to the lowest variance is less than 2, then there are negligible fluctuations in the analysis of variance. If there is a question about equality of variance, Bartlett's test will be useful. Details of this test can be found in Montgomery (1991).

- The observations of the ijkth cell are a random sample of size n from a normally distributed population with mean μ_{ijk} and variance σ^2.

- Input variables or factors must be independent of each other.

- The experiment is run in a random order. Randomization can be accomplished by using a random number table or other technique.

- The conclusions of a fixed-effects model apply only to the levels of the factors included in the experiment. The conclusions cannot be extended to include any level not explicitly in the experiment.

5 APPLICATION CONSIDERATIONS

- If there are more than two or three variables in an analysis of variance, it is easier to use a computer program. Several analysis of variance software packages exist for personal computers. Analysis of variance routines can be found in statistical packages as well as design of experiments packages. The sheer number of computations goes up geometrically with the number of variables added to the analysis of variance, and without the use of a software package, it is very easy to make an arithmetic error.

- For single replicated designs (designs where there is only one observation per experimental combination) there are no degrees of freedom in which to compute an error sum of squares (*SSE*). These problems can be handled with Daniel's Normal Probability Plot. For further details refer to Montgomery (1991).

- If all of the factors are at only two levels, a high and a low level, then the manual computations can be greatly reduced by using Yates' Algorithm. For further details refer to any of the references. Yates also has an algorithm if all of the factors are at three levels.

- While equality of variance is an assumption in the analysis of variance, variances can and should be checked with residual plots. Residual plots of the factors, residual plots of the experiment over time, and residual plots of the actual observed values versus the predicted values should be checked for outliers as well as any patterns. These residual plots should be structureless. A normal probability plot of the residuals is also recommended to check the normality assumption about the residuals. Good computer programs will have the ability to generate these plots easily.

- Coding observations can simplify the calculations and improve the accuracy in the analysis of variance. Basic operations of addition, subtraction, multiplication, and division can be performed on each of the individual observations of an experiment without changing the resulting F-ratios of the analysis of variance. For example, if all of the observations of an experiment are of the form $0.xx$, then they can all be multiplied by 100 to remove the decimal point. This transformation will make the numbers easier to use in the computations and will help to reduce possible round-off errors. The F-ratios for the coded data will be the same as the F-ratios for the uncoded data. It is not generally necessary to code the data if a computer program is being used.

- Data transformations can be used to stabilize the observations of an experiment with respect to variance. Some common transformations, in order of strength, are the square root, the logarithmic, the reciprocal square root, and the reciprocal. Poisson or count data are best handled with a square-root transformation. For more information on data transformations in the analysis of variance, refer to Montgomery (1991).

6 APPLICATION SCENARIOS

The data in Figure 3.1 will be referred to in each of the following applications scenarios as well as the application roadmap. The data are supplied for the sake of illustration. Assume that the data are in coded form, do not try to associate specific units or meaning to numbers. Also assume for the sake of illustration that the sampling size is appropriate and was taken in a random fashion.

Electrical Engineering: A circuit designer for Gizmos, Inc., decided to evaluate the different manufacturers of three components that are used in one of her applications. The components are the bipolar transistor, diode, and FET. There are two different manufacturers of the transistor, three manufacturers of the diode, and two of the FET. She designed a replicated two-by-three-by-two factorial experiment. Her response was the overall voltage drop across the circuit board. The engineer performed an N-way analysis of variance to

	$C1$		$C2$	
	$A1$	$A2$	$A1$	$A2$
$B1$	6	32	24	43
	0	36	26	46
$B2$	25	24	15	36
	12	16	21	30
$B3$	15	14	16	25
	6	21	24	38

FIGURE 3.1. Coded Data for Application Scenarios

analyze her results. The resulting data are displayed in Figure 3.1.

Comment: The independent variables are the two transistor manufacturers, A, the three diode manufacturers, B, and the two FET manufacturers, C. The dependent variable is the voltage drop.

Mechanical Engineering: A mechanical engineer from Sili Semiconductor was developing a second-generation plastic package for an integrated circuit. This second-generation package needed to be more robust to mechanical shock and stress. The engineer chose to evaluate two heat-sink designs, three mold compounds, and two solder types. The experiment that he designed was a two-by-three-by-two factorial design. For an estimate of error, he replicated the experiment once. By using a jig that he created, the engineer was able to stress the part with a press while measuring an electrical output parameter on a meter. His response was the amount of stress that could be withstood by the IC prior to the package giving way. The engineer used an N-way ANOVA to perform the analysis. The resulting data are displayed in Figure 3.1.

Comment: The independent variables are the two lead frame designs, A, the three mold compounds, B, and the two solder types, C. The dependent variable is the mechanical stress.

Process Engineering: A photo resist engineer at Elmo Semiconductor wished to evaluate two development solutions and three different resists as well as the use of a pre-bake adhesion promoter. She was interested in the resolution of the resist after development. The engineer devised a grading scale for the wafers such that the higher the score, the better the wafer was. She designed a replicated two-by-three-by-two factorial experiment and would use an N-way analysis of variance to analyze the results. The resulting data are displayed in Figure 3.1.

Comment: The independent variables are the two development solutions, A, the three resist types, B, and the adhesion promoter, with or without adhesion promoter, C. The dependent variable is the resolution.

Software Engineering: Groovey Graphics, Inc., a graphics design company, was interested in upgrading its workstations to a more flexible alternative that would allow a designer to use the terminal as a personal computer as well as a mainframe terminal. To accomplish this, Groovey Graphics hired a software engineering consultant to run a study to determine which of the available options would offer the quickest graphics refresh time. The engineer decided to try two types of modems, Alpha and Beta, three workstations, and two emulator packages. The engineer knew that his experimental design was a two-by-three-by-two factorial design, and that he would need to use an N-way analysis of variance to analyze his results. For an estimate of error, he chose to replicate the experiment once. The resulting data are displayed in Figure 3.1.

Comment: The independent variables are the two types of modems, A, the three workstations, B, and the two emulator packages, C. The dependent variable is the graphics refresh time.

Manufacturing: A wafer manufacturing engineer at Farkle Semicon received funding from upper management to convert her diffusion operation's wafer handling from manual to automated transfer equipment. Rather than arbitrarily purchasing new transfer equipment, quartz boats, and wafer cassettes, she chose to evaluate her options by requesting from the vendors a one-month free evaluation of their equipment. With her technical expertise, she was able to narrow the potential candidates down to two automated wafer transfer systems, three different quartz boat designs, and two different wafer cassettes. The engineer designed a two-by-three-by-two factorial experiment. She measured the number of loading errors per 100 transfer cycles for each experimental cell. A transfer cycle consists of unloading from the cassette, loading into a quartz boat, unloading from the quartz boat, and reloading into the cassette. She replicated the experiment once to establish an estimate of error. She used an N-way analysis of variance to analyze her results. The resulting data are displayed in Figure 3.1.

Comment: The independent variables are the two types of automated transfer equipment, A, the three quartz boat designs, B, and the two wafer cassettes, C. The dependent variable is the errors per 100 transfer cycles.

Administration: The manager of an administrative office at General Distribution, Inc., decided to do something about all of the complaints his office had been receiving from other divisions at GD. The complaints were concerning the clarity of the many 11-by-14-inch computer printouts that are created by his organization. His goal was to resolve the problem without replacing the costly FMC8000B line printers that were currently being used. Instead, he chose to design and run a two-by-three-by-two factorial experiment. The factors that he wanted to evaluate were two types of ink, three types of computer paper, and two different ink jets. Each run in the experiment consisted of running the same lengthy sales report for GD's western region. The clarity of the report was recorded based on the number of illegible characters in the report. He replicated the experiment once so as to have an estimate of error. The manager used an N-way ANOVA to analyze his experiment. The resulting data are displayed in Figure 3.1.

Comment: The independent variables are the two types of printer ink, A, the three types of computer paper, B, and the two different ink jets, C. The dependent variable is the number of illegible characters in the sales report.

Facilities: Shinola Metals, Inc., is a major manufacturer of stamped sheet metal products used in the construction industry. Due to health risks near the finance and administration offices at one of Shinola's facilities, the cafeteria needed to be moved to a more isolated part of the factory. In its place, a stamping operation was to be established. However, this type of manufacturing is extremely noisy and would distract the finance and administration employees a great deal, especially during their breaks. To minimize the noise level in the adjacent offices, a facilities engineer decided to evaluate different noise reduction, or insulating, materials for the ceiling, walls, and floor of the new stamping area. After an examination of the advertised properties of the available materials, the engineer chose to study the noise reduction effects of two ceiling materials, three wall materials, and two floor materials. To do this, she assembled a small model of the stamping area with removable walls, floor, and ceiling. She put a tape player inside the model with the volume set on high while playing an AC/DC song, "Dirty Deeds." A noise detector was placed 20 inches from the model and recorded the average noise level that escaped the model. The experimental design she used was a replicated two-by-three-by-two factorial design. She analyzed her data with an N-way analysis of variance. The resulting data are displayed in Figure 3.1.

Comment: The independent variables are two ceiling materials, A, three wall materials, B, and two floor materials, C. The dependent variable is the average noise level.

Finance: After evaluating the Pareto of errors in the latest month-end financial report, the vice president of finance at SOL Communications, Inc., decided to put an end to the high level of simple mathematical errors that were plaguing the report. The mainframe computer system used by the company is not capable of performing all of the calculations required, including the majority of the smaller sales offices' monthly tax reports. He personally chose a newly graduated accountant to focus on this problem full time until it was resolved. The new accountant, with the help of others in her office, drew up a fishbone diagram of all of the potential causes for math errors. She spent the following two weeks paring down the suspects until she identified the biggest culprit—the calculators. The calculators being used were old, clunky adding machines. She decided to evaluate the options offered by the major calculator company, Kansas Instruments, Inc. The accountant designed a two-by-three-by-two factorial experiment and ordered a sample of calculators that included the following options: LED versus LCD displays; small, medium, and large buttons; and small and large displays. She got the participation of the three other accountants in the office and had them each enter and perform the calculations for one of the month-end data entry forms, using each of the twelve different calculators. She summed up all of the errors of the three accountants for each calculator that they used. She replicated the experiment once for a total of 24 runs. The accountant analyzed the data using analysis of variance. The resulting data are displayed in Figure 3.1.

Comment: The independent variables are the LED and LCD displays, A, the three button sizes, B, and the two display sizes, C. The dependent variable is the total number of errors by the three accountants per experimental run.

Personnel: The annual employee survey at Healthy Desserts, Inc., producer of best-selling desserts such as Brussels Sprout Swirl Ice Cream, showed a strong employee dissatisfaction with the way the personnel department was handling the employees profit-sharing requests. Upon further investigation, it was found that the changes the employees were making to their profit-sharing programs and salary-investment program accounts were not being properly entered by personnel. Employees must complete a form, in triplicate, stating the changes that they wish to have made to their payroll deductions for profit-sharing and salary investments. The original is put in the employees files, while the carbon copies are sent to payroll and the Healthy Desserts Credit Union. Apparently, the carbon copies were not clear enough for proper interpretation. Therefore, the manager of personnel decided to run a two-by-three-by-two replicated factorial experiment. He wished to evaluate two types of carbon material; medium, fine, and extra fine pens; and two types of paper. He asked two employees to write out every alphabetic and numeric character, with each character being approximately one-quarter inch in size. The total number of unidentifiable characters were recorded for each cell of the experiment. N-way analysis of variance was used to analyze the recorded data. The resulting data are displayed in Figure 3.1.

Comment: The independent variables are the two carbon materials, A, the three pen sizes, B, and the two paper types, C. The dependent variable is the total number of errors by the three accountants per experimental run.

Sales: A marketer for a major tool manufacturer, Speed Wrench, Inc., was given the task of developing a new novelty item with the company logo on it for gifts to prospective buyers at the upcoming trade shows. Due to popular demand, she chose to go with a pocket knife with the logo printed on the side. However, due to cost constraints, she could not put both a screw driver and a cork screw in the same knife. She also did not know if she should have the knives colored red, blue, or green. The manufacturer also offered the knives with a key chain attachment or a case. Again, she could not decide. What she did decide on was to run an experiment containing the above variables. The design that she ran was a two-by-three-by-two factorial experiment that she replicated once by having the samples available during two days of the next trade show. She would display the samples at the beginning of each day, and record at the end of the day the number of knives taken from each sample. She used an N-way ANOVA to analyze her data. The resulting data are displayed in Figure 3.1.

Comment: The independent variables are the two internal accessories (screw driver or cork screw), A, the three colors, B, and the two external accessories (case or key ring), C. The dependent variable is the number of knives taken by customers.

7 APPLICATION ROADMAP

Step 1. State the risk level:

$\alpha = 0.05$ or, equivalently, $\alpha = 5\%$.

Step 2. Computations: Compute the factor and cell totals.

Step 2.1. Compute the totals for each treatment of factor A ($y_{i...}$) and record in the data table:

At the low level of C ($C1$):
$y_{1...} = 6 + 0 + 25 + 12 + 15 + 6 = 64$.
At the high level of C ($C2$):
$y_{1...} = 24 + 26 + 15 + 21 + 16 + 24 = 126$.
$y_{1...} = 64 + 126 = 190$.

At the low level of C ($C1$):
$y_{2...} = 32 + 36 + 24 + 16 + 14 + 21 = 143$.
At the high level of C ($C2$):
$y_{2...} = 43 + 46 + 36 + 30 + 25 + 38 = 218$.
$y_{2...} = 143 + 218 = 361$.

Step 2.2. Compute the totals for each treatment of factor B ($y_{.j.}$) and record in the data table:

$$y_{.1..} = 6 + 0 + 32 + 36 + 24 + 26$$
$$+ 43 + 46 = 213,$$
$$y_{.2..} = 25 + 12 + 24 + 16 + 15 + 21$$
$$+ 36 + 30 = 179,$$
$$y_{.3..} = 15 + 6 + 14 + 21 + 16 + 24$$
$$+ 25 + 38 = 159.$$

Step 2.3. Compute the totals for each treatment of factor C ($y_{..k}$) and record in the data table:

$$y_{..1} = 6 + 0 + 32 + 36 + 25 + 12 + 24 + 16$$
$$+ 15 + 6 + 14 + 21 = 207.$$
$$y_{..2} = 24 + 26 + 43 + 46 + 15 + 21 + 36$$
$$+ 30 + 16 + 24 + 25 + 38 = 344.$$

Step 2.4. Compute the total for each cell (y_{ijk}) and record this value in a circle next to the corresponding cell as shown in Figure 3.2:

$$y_{111.} = 6 + 0 = 6,$$
$$y_{112.} = 24 + 26 = 50,$$
$$y_{121.} = 25 + 12 = 37,$$
$$y_{122.} = 15 + 21 = 36,$$
$$y_{131.} = 15 + 6 = 21,$$
$$y_{132.} = 16 + 24 = 40,$$
$$y_{211.} = 32 + 36 = 68,$$
$$y_{212.} = 43 + 46 = 89,$$
$$y_{221.} = 24 + 16 = 40,$$
$$y_{222.} = 36 + 30 = 66,$$
$$y_{231.} = 14 + 21 = 35,$$
$$y_{232.} = 25 + 38 = 63.$$

| | C1 | | C2 | | |
	A1	A2	A1	A2	$y_{.i..}$
B1	6 ⑥ 0	32 ㉞ 36	24 ㊿ 26	43 ⑧⑨ 46	213
B2	25 ㊲ 12	24 ㊵ 16	15 ㊱ 21	36 ㊻ 30	179
B3	15 ㉑ 6	14 ㉟ 21	16 ㊵ 24	25 ㊿ 38	159
$y_{i..}$	64	143	126	218	$y_{....}$
	$y_{1..}$ 190		$y_{2..}$ 361		551
$y_{..k.}$	207		344		

FIGURE 3.2. Coded Data With Factor and Cell Totals

Step 2.5. Compute the grand total ($y....$). One way is to add the B treatment totals and record the grand total at the bottom of that column.

Using the totals of B:
$$y.... = 213 + 179 + 159 = 551.$$

Step 3. Construct the ANOVA table. Replace the entries in Table 3.1 with the computed values.

Step 4. Compute the degrees of freedom.

Step 4.1. Compute the degrees of freedom for treatment A and record in the ANOVA table:

degrees of freedom (A) = number of A treatments $- 1 = a - 1$.

degrees of freedom (A) $= 2 - 1 = 1$.

Step 4.2. Compute the degrees of freedom for treatment B and record in the ANOVA table:

degrees of freedom (B) = number of B treatments $- 1 = b - 1$.

degrees of freedom (B) $= 3 - 1 = 2$.

Step 4.3. Compute the degrees of freedom for the AB interaction and record in the ANOVA table:

degrees of freedom (AB) $= (a - 1)(b - 1)$.

degrees of freedom (AB) $= (2 - 1)(3 - 1)$
$= 1 \times 2 = 2$.

Step 4.4. Compute the degrees of freedom for treatment C and record in the ANOVA table:

degrees of freedom (C) = number of C treatments $- 1 = c - 1$.

degrees of freedom (C) $= 2 - 1 = 1$.

Step 4.5. Compute the degrees of freedom for the AC interaction and record in the ANOVA table:

degrees of freedom (AC) $= (a - 1)(c - 1)$.

degrees of freedom (AC) $= (2 - 1)(2 - 1)$
$= 1 \times 1 = 1$.

TABLE 3.1 *General ANOVA Table*

Source of Variation	Degrees of Freedom	Sum of Squares	Mean Square	F
Factor A	$a-1$	SSA	$MSA = \dfrac{SSA}{(a-1)}$	$\dfrac{MSA}{MSE}$
Factor B	$b-1$	SSB	$MSB = \dfrac{SSB}{(b-1)}$	$\dfrac{MSB}{MSE}$
Interaction AB	$(a-1)(b-1)$	SSAB	$MSAB = \dfrac{SSAB}{(a-1)(b-1)}$	$\dfrac{MSAB}{MSE}$
Factor C	$c-1$	SSC	$MSC = \dfrac{SSC}{(c-1)}$	$\dfrac{MSC}{MSE}$
Interaction AC	$(a-1)(c-1)$	SSAC	$MSAC = \dfrac{SSAC}{(a-1)(c-1)}$	$\dfrac{MSAC}{MSE}$
Interaction BC	$(b-1)(c-1)$	SSBC	$MSBC = \dfrac{SSBC}{(b-1)(c-1)}$	$\dfrac{MSBC}{MSE}$
Interaction ABC	$(a-1)(b-1)(c-1)$	SSABC	$MSABC = \dfrac{SSABC}{(a-1)(b-1)(c-1)}$	$\dfrac{MSABC}{MSE}$
Error	$ab(n-1)$	SSE	$MSE = \dfrac{SSE}{ab(n-1)}$	
Total	$abn-1$	SST		

Step 4.6. Compute the degrees of freedom for the BC interaction and record in the ANOVA table:

$$\text{degrees of freedom } (BC) = (b-1)(c-1).$$
$$\text{degrees of freedom } (BC) = (3-1)(2-1)$$
$$= 2 \times 1 = 2.$$

Step 4.7. Compute the degrees of freedom for the ABC interaction and record in the ANOVA table:

$$\text{degrees of freedom } (ABC) = (a-1)(b-1)$$
$$(c-1).$$
$$\text{degrees of freedom } (ABC) = (2-1)(3-1)$$
$$(2-1)$$
$$= 1 \times 2 \times 1 = 2.$$

Step 4.8. Compute the degrees of freedom for error and record in the ANOVA table:

$$\text{degrees of freedom (error)} = abc \text{ (number of}$$
$$\text{replicates} - 1)$$
$$= abc(n-1).$$
$$\text{degrees of freedom (error)} = (2 \times 3 \times 2)$$
$$(2-1)$$
$$= 12 \times 1 = 12.$$

Step 4.9. Compute the degrees of freedom for the total and record in the ANOVA table. Add the degrees of freedom for the treatments, blocks, and error. This sum should equal the degrees of freedom for the total.

$$\text{degrees of freedom (total)} = \text{total number of}$$
$$\text{observations}$$
$$- 1 = abcn - 1.$$
$$\text{degrees of freedom (total)} = (2 \times 3 \times 2 \times 2)$$
$$- 1 = 24 - 1 = 23.$$

Step 5. Compute the sum of squares.

Step 5.1. Compute the correction factor:

$$C = \frac{y^2...}{abcn} = \frac{551^2}{24} = 12650.04167.$$

Step 5.2. Compute the sum of squares total (SST) and record in the ANOVA table:

$$SST = \sum_{i=1}^{a} \sum_{j=1}^{b} \sum_{k=1}^{c} \sum_{l=1}^{n} y_{ijkl}^2 - C,$$

$$\begin{aligned} SST = (6^2 &+ 0^2 + 25^2 + 12^2 + 15^2 + 6^2 + 32^2 \\ &+ 36^2 + 24^2 + 16^2 + 14^2 + 21^2 \\ &+ 24^2 + 26^2 + 15^2 + 21^2 \\ &+ 16^2 + 24^2 + 43^2 + 46^2 + 36^2 \\ &+ 30^2 + 25^2 + 28^2) - C, \end{aligned}$$

$$SST = 15835 - 12650.04167 = 3184.95833.$$

Step 5.3. Compute the sum of squares for the A treatments (SSA) and record in the ANOVA table:

$$SSA = \sum_{i=1}^{a} \frac{y_{i...}^2}{bcn} - C,$$

$$SSA = \frac{(190^2 + 361^2)}{12} - C,$$

$$SSA = \frac{166421}{12} - 12650.04167 = 13868.41667$$
$$- 12650.04167 = 1218.375.$$

Step 5.4. Compute the sum of squares for the B treatments (SSB) and record in the ANOVA table:

$$SSB = \sum_{j=1}^{b} \frac{y_{.j..}^2}{acn} - C,$$

$$SSA = \frac{(213^2 + 179^2 + 159^2)}{8} - C,$$

$$SSA = \frac{102691}{8} - 12650.04167$$
$$= 12836.375 - 12650.04167$$
$$= 186.33333.$$

Step 5.5. Compute the sum of squares of the interaction $AB(SSAB)$ and record in the ANOVA table:

$$SSAB = \sum_{i=1}^{a} \sum_{j=1}^{b} \frac{y_{ij..}^2}{cn} - C - SSA - SSB,$$

$$SSAB = \frac{\left[(6 + 50)^2 + (37 + 36)^2 + (21 + 40)^2 + (68 + 89)^2 \atop + (40 + 66)^2 + (35 + 63)^2\right]}{4}$$

$$- SSA - SSB - C$$

$$SSAB = \frac{(56^2 + 73^2 + 61^2 + 157^2 + 106^2 + 98^2)}{4}$$

$$- SSA - SSB - C,$$

$$SSAB = \frac{57675}{4} - 1218.375 - 186.33333 - 12650.04167,$$

$$SSAB = 14418.75 - 1218.375 - 186.33333 - 12650.04167$$
$$= 364.00000.$$

Step 5.6. Compute the sum of squares for the C treatments (SSC) and record in the ANOVA table:

$$SSC = \sum_{k=1}^{c} \frac{y_{y..k.}^2}{abn} - C,$$

$$SSC = \frac{(207^2 + 344^2)}{12} - C,$$

$$SSC = \frac{161185}{12} - 12650.04167 = 13432.08333$$

$$- 12650.04167 = 782.04166.$$

Step 5.7. Compute the sum of squares of the interaction $AC(SSAC)$ and record in the ANOVA table:

$$SSAC = \sum_{i=1}^{a} \sum_{k=1}^{c} \frac{y_{i.k.}^2}{bn} - C - SSA - SSC,$$

$$SSAC = \cfrac{\begin{bmatrix}(6 + 37 + 21)^2 + (68 + 40 + 35)^2 + (50 + 36 + 40)^2 \\ + (89 + 66 + 63)^2\end{bmatrix}}{6}$$
$$- SSA - SSC - C$$

$$SSAC = \frac{(64^2 + 143^2 + 126^2 + 218^2)}{6} - SSA - SSC - C,$$

$$SSAC = \frac{87945}{6} - 1218.375 - 782.04166 - 12650.04167,$$

$$SSAC = 14657.5 - 1218.375 - 782.04166 - 12650.04167$$
$$= 7.04167.$$

Step 5.8. Compute the sum of squares of the interaction $BC(SSBC)$ and record in the ANOVA table:

$$SSBC = \sum_{j=1}^{b} \sum_{k=1}^{c} \frac{y_{.jk.}^2}{an} - C - SSB - SSC,$$

$$SSBC = \cfrac{\begin{bmatrix}(6 + 68)^2 + (50 + 89)^2 + (37 + 40)^2 + (36 + 66)^2 \\ + (21 + 35)^2 + (40 + 63)^2\end{bmatrix}}{4}$$
$$- SSB - SSC - C,$$

$$SSBC = \frac{(74^2 + 139^2 + 77^2 + 102^2 + 56^2 + 103^2)}{4}$$
$$- SSB - SSC - C,$$

$$SSBC = \frac{54875}{4} - 186.33333 - 782.04166 - 12650.04167,$$

$$SSBC = 13718.75 - 186.33333 - 782.04166 - 12650.04167$$
$$= 100.33334.$$

Step 5.9. Compute the sum of squares of the interaction $ABC(SSABC)$ and record in the ANOVA table:

$$SSABC = \sum_{i=1}^{a} \sum_{j=1}^{b} \sum_{k=1}^{c} \frac{y_{ijk.}^2}{n} - C - SSA - SSB - SSAB$$
$$- SSC - SSAC - SSBC,$$

$$SSABC = \frac{(6^2 + 68^2 + 50^2 + 89^2 + 37^2 + 40^2 + 36^2 + 66^2 + 21^2 + 35^2 + 40^2 + 63^2)}{2}$$

$$- SSA - SSB - SSAB - SSC - SSAC - SSBC - C,$$

$$SSABC = \frac{30937}{2} - 1218.375 - 186.33333$$

$$- 364.00000 - 782.04166 - 7.04167 - 100.33334$$
$$- 12650.04167,$$

$$SSABC = 15468.5 - 1218.375 - 186.33333$$

$$- 364.00000 - 782.04166 - 7.04167 - 100.33334$$
$$- 12650.04167 = 160.33333.$$

Step 5.10. Compute the sum of squares for the error (SSE) by subtraction and record in the ANOVA table:

$$SSE = SST - SSA - SSB - SSAB - SSC - SSAC - SSBC - SSABC,$$

$$SSE = 3184.95833 - 1218.375 - 186.33333$$
$$- 364 - 782.04166 - 7.04167$$
$$- 100.33334 - 160.33333,$$

$$SSE = 366.5.$$

Step 6. Compute mean squares.
Step 6.1. Compute the mean square treatment $A(MSA)$ and record in the ANOVA table:

$$MSA = \frac{SSA}{\text{degrees of freedom for treatment } A}$$

$$= \frac{1218.375}{1} = 1218.375.$$

Step 6.2. Compute the mean square of treatment $B(MSB)$ and record in the ANOVA table:

$$MSB = \frac{SSB}{\text{degrees of freedom for treatment } B}$$

$$= \frac{186.33333}{2} = 93.16667.$$

Step 6.3. Compute the mean square of interaction $AB(MSAB)$ and record in the ANOVA table:

$$MSAB = \frac{SSAB}{\text{degrees of freedom for interaction } AB}$$

$$= \frac{364}{2} = 182.$$

Step 6.4. Compute the mean square of treatment $C(MSC)$ and record in the ANOVA table:

$$MSC = \frac{SSC}{\text{degrees of freedom for treatment } C}$$

$$= \frac{782.04166}{1} = 782.04166.$$

Step 6.5. Compute the mean square of treatment $AC(MSAC)$ and record in the ANOVA table:

$$MSAC = \frac{SSAC}{\text{degrees of freedom for interaction } AC}$$

$$= \frac{7.04167}{1} = 7.04167.$$

Step 6.6 Compute the mean square of interaction $BC(MSBC)$ and record in the ANOVA table:

$$MSBC = \frac{SSBC}{\text{degrees of freedom for interaction } BC}$$

$$= \frac{100.33333}{2} = 50.16667.$$

Step 6.7. Compute the mean square of interaction ABC $(MSABC)$ and record in the ANOVA table:

$$MSABC = \frac{SSABC}{\text{degrees of freedom for interaction } ABC}$$

$$= \frac{160.33333}{2} = 80.16667.$$

Step 6.8. Compute the mean square error (MSE) and record in the ANOVA table:

$$MSE = \frac{SSE}{\text{degrees of freedom for error}}$$

$$= \frac{366.5}{12} = 30.54167.$$

Step 7. Compute F-Values.

Step 7.1. Compute the F-value for treatment A and record in ANOVA table:

$$F(A) = \frac{MSA}{MSE} = \frac{1218.375}{30.54167} = 39.89222.$$

Step 7.2. Compute the F-value for treatment B and record in ANOVA table:

$$F(B) = \frac{MSB}{MSE} = \frac{93.16667}{30.54167} = 3.05048.$$

Step 7.3. Compute the F-Value for interaction AB and record in ANOVA table:

$$F(AB) = \frac{MSAB}{MSE} = \frac{182}{30.54167} = 5.95907.$$

Step 7.4. Compute the F-value for treatment C and record in ANOVA table:

$$F(C) = \frac{MSC}{MSE} = \frac{782.04166}{30.54167} = 25.60573.$$

Step 7.5. Compute the F-value for interaction AC and record in ANOVA table:

$$F(AC) = \frac{MSAC}{MSE} = \frac{7.04167}{30.54167} = 0.23056.$$

Step 7.6. Compute the F-value for interaction BC and record in ANOVA table:

$$F(BC) = \frac{MSBC}{MSE} = \frac{50.16667}{30.54167} = 1.64256.$$

Step 7.7. Compute the F-value for interaction ABC and record in ANOVA table. After this step the ANOVA table should look like Table 3.2. The numbers in the ANOVA table have been rounded to three significant figures.

$$\text{F}(ABC) = \frac{MSABC}{MSE} = \frac{3.1508}{30.54167} = 2.62483.$$

Step 8. Determine the critical F-regions. F-tables can be found in all the references as well as most statistics books.

Step 8.1. Determine the critical F-region for the A treatments. Look up $F_{\alpha,\nu1,\nu2}$ in an F-table, where α is the risk factor, ν_1 is the number of degrees of freedom for the A treatments, and ν_2 is the number of degrees of freedom for the error. Since factor C and the AC interaction also have 1 degree of freedom, the critical F-region is the same as for factor A:

$$F_{\alpha,\nu1,\nu2} = F_{0.05,1,12} = 4.75.$$

Step 8.2. Determine the critical F-region for the B treatments. Look up $F_{\alpha,\nu1,\nu2}$ in an F-table, where α is the risk factor, ν_1 is the number of degrees of

TABLE 3.2 *Complete ANOVA Table for Scenario Data*

Source of Variation	Degrees of Freedom	Sum of Squares	Mean Square	F
A	1	1218.375	1218.375	39.892*
B	2	186.333	93.167	3.050
AB	2	364.000	182.000	5.959*
C	1	782.042	782.042	25.606*
AC	1	7.042	7.042	0.231
BC	2	100.333	50.167	1.643
ABC	2	160.333	80.167	2.625
Error	12	366.500	30.542	
Total	23	3184.958		

freedom for the B treatments, and ν_2 is the number of degrees of freedom for the error. Since the AB interaction, the BC interaction and the ABC interaction also have two degrees of freedom, the critical F-region is the same as for factor B.

$$F_{\alpha,\nu1,\nu2} = F_{0.05,2,12} = 3.89.$$

Step 9. Draw Conclusions: Compare the calculated F-value to the critical F-value for each of the main effects and interactions. The significant effects are those where the calculated F-value is greater than the critical F-value. In this example,

$$F(A) = 39.89222 > F_{0.05,1,12} = 4.75,$$

$$F(AB) = 4.95907 > F_{0.05,2,12} = 3.89, \text{ and}$$

$$F(C) = 25.60573 > F_{0.05,1,12} = 4.75.$$

The F-values of the remaining effects are less than their corresponding critical F-values. The significant effects for this example are A, C, and the AB interaction. Since the AB interaction was found to be significant, it is useful to view this interaction graphically.

The graph of the AB interaction shows that at the low level of A there is not much difference in the output no matter which level of B is chosen. At the high level of A, there is much variation in the output variable, depending on the level of factor B. The interaction plot shows that the $B2$ and $B3$ lines are parallel and relatively close together. This suggests that there is little difference in the output between levels 2 and 3 of factor B. The experimenter should use the information in an interaction plot to advantage. For example, if the desired response is low, then the experimenter can choose the cheapest level of factor B to operate the process.

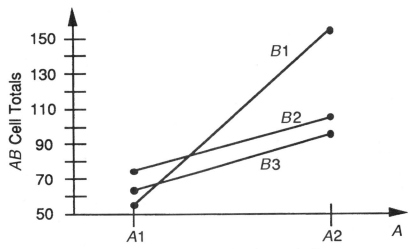

FIGURE 3.3. *AB* **Interaction Plot for Scenario Data**

The experimenter also should check model adequacy with residual plots.

REFERENCES AND BIBLIOGRAPHY

Hicks, C. R. 1964. *Fundamental Concepts in the Design of Experiments*. New York: Holt, Rinehart & Winston.

Hines, W. W., and D. C. Montgomery. 1991. *Probability and Statistics in Engineering and Management Science*, 3rd Edition. New York: John Wiley and Sons.

Miller, I., and J. E. Freund. 1985. *Probability and Statistics for Engineers*, 3rd Edition. Englewood Cliffs, New Jersey: Prentice-Hall.

Montgomery, D. C. 1991. *Design and Analysis of Experiments*, 3rd Edition. New York: John Wiley and Sons.

Walpole, R. E., and R. H. Myers. *Probability and Statistics for Engineers and Scientists*, 4th Edition. New York: MacMillan.

CHAPTER 4

Randomized Blocks

1 DESCRIPTION

The analysis of variance is a technique performed to determine if there is a statistical difference between the levels of the factor of interest. The randomized block design is used when an experimental design has systematically restricted the randomization in order to control the variability from a nuisance variable. The main focus of the randomized block design is on a single factor, with the second factor, the block, being the nuisance factor. The randomized block design is an extension of the paired t-test. Like a paired t-test, the randomized block design reduces the noise in an experiment by blocking the effects of the nuisance variable. But, in the case of the randomized block design, we are concerned with more than two levels of the factor of interest. The analysis of variance for this design now partitions the variability into a source of variability for treatments, a source of variability for blocks, and a source of variability for error.

2   LIST OF EQUATIONS

Sum of Squares Identity for a Randomized Block

$$\sum_{i=1}^{a} \sum_{j=1}^{b} \left(y_{ij} - \overline{y}.. \right)^2 = b \sum_{i=1}^{a} \left(\overline{y}_{i.} - \overline{y}.. \right)^2 + a \sum_{j=1}^{b} \left(\overline{y}_{.j} - \overline{y}.. \right)^2$$

$$+ \sum_{i=1}^{a} \sum_{j=1}^{b} \left(y_{ij} - \overline{y}_{i.} - \overline{y}_{.j} - \overline{y}.. \right)^2,$$

where

y_{ij} = an observation in the ith treatment and the jth block,

$\overline{y}..$ = the mean of all ab observations,

$\overline{y}_{i.}$ = the mean of the observations for the ith treatment,

$\overline{y}_{.j}$ = the mean of the observations for the jth block,

a = the number of levels of treatment A, and

b = the number of blocks.

3   VARIABLES

- The input or independent variable is called a *factor* or a *treatment*. The independent variable is controlled by the experimenter, for fixed-effects models. The independent variable is a *categorical variable*.

- The dependent variable is called a *response variable*. The response variable is the measured or the observed value from the experiment.

- The independent variables may be discrete or continuous. When discrete, the factor is separated naturally into levels. For example, two different mold compounds have two natural levels or groups. Continuous variables (i.e., temperature, pressure, time. etc.) need to have levels established by the experimenter.

- The blocks refer to a source of variation that is not the focus of the experiment; however, is identifiable and is worth removing from the error term. The blocks are often referred to as nuisance variables. Common blocks are measurement equipment, operators, time, etc.

4 UNDERLYING ASSUMPTIONS

- The errors of the linear model associated with the analysis of variance are normally and independently distributed with a mean of zero and a constant variance.

- The variance of the response is assumed to be constant for all levels of a given factor. The analysis of variance is robust to mild violations of equivalence of variance. As a rule of thumb, if the ratio of the highest variance to the lowest variance is less than 2, then there are negligible fluctuations in the analysis of variance. If there is a question about equality of variance, Bartlett's test will be useful. Details of this test can be found in Montgomery (1991).

- The observations of the ijth cell are a random sample of size n from a normally distributed population with mean μ_{ij} and variance σ^2.

- Input variables or factors must be independent of each other.

- The experiment is run in a random order. The observations of each block should be randomized. Randomization can be accomplished by using a random number table or other technique.

5 APPLICATION CONSIDERATIONS

- Several analysis of variance software packages exist for personal computers. Analysis of variance routines can be found in statistical packages as well as design of experiments packages. The sheer number of computations goes up geometrically with the number of variables added to the analysis of variance. Without the use of a software package, it is very easy to make an arithmetic error.

- While equality of variance is an assumption in the analysis of variance, variances can and should be checked with residual plots. Residual plots of the factors, residual plots of the experiment over time, and residual plots of the actual observed values versus the predicted values should be checked for outliers as well as any patterns. These residual plots should be structureless. A normal probability plot of the residuals is also recommended to check the normality assumption about the residuals. Good computer programs will have the ability to generate these plots easily.

- Coding observations can simplify the calculations and improve the accuracy in the analysis of variance. Basic operations of addition, subtraction, multiplication, and division can be performed on each of the individual observations of an experiment without changing the resulting F-ratios of the analysis of variance. For example, if all of the observations of an experiment are of the form $0.xx$, then they can all be multiplied by 100 to remove the decimal point. This transformation will make the numbers easier to use in the computations and help reduce possible round-off errors. The F-ratios for the coded data will be the same as the F-ratios for the uncoded data. It is not generally necessary to code the data if a computer program is being used.

- Data transformations can be used to stabilize the observations of an experiment with respect to variance. Some common transformations in order of strength are the square root, the logarithmic, the reciprocal square root, and the reciprocal. Poisson or count data are best handled with a square root transformation. For more information on data transformations in the analysis of variance, refer to Montgomery (1991).

6 APPLICATION SCENARIOS

The data in Figure 4.1 will be referred to in each of the following applications scenarios as well as the application roadmap. The data are supplied for the sake of illustration. Assume that the data are in coded form; do not try to associate specific units or meaning to numbers. Also assume for the sake of illustration that the sampling size is appropriate and was taken in a random fashion.

Electrical Engineering: An electrical engineer at Wizmo Electronics wished to evaluate four different suppliers, A, B, C, and D, for a particular transistor that he is currently using in one of Wizmo's products. Each of the suppliers is very capable of delivering product that is within the required tolerance; however, the engineer was interested in determining which brand degrades the least under harsh environmental conditions. To do this, the engineer needed to insert the units to be tested into a test board that would be placed in an environmental chamber. The engineer realized that the test board could not accommodate a sufficient number of parts to satisfy a statistically sound sample. In fact, this experiment would require four test boards. After completing initial readouts on the transistors to be tested, the engineer divided up the units for each supplier into four groups, one group for each board, thus allowing each supplier to be tested on each of the four boards. The transistors were all randomly inserted into the sockets on the boards and placed into the environmental chamber for

| | Blocks | | | |
Treatments	A	B	C	D
1	25	33	36	23
2	23	22	11	17
3	9	6	3	12
4	19	20	22	29

FIGURE 4.1. Coded Scenario Data

1000 hours. After receiving the 1000 hours of stress, the transistors were measured once again and the degradation determined. The average degradation was determined for each of the groups and analyzed using a randomized block design. The resulting data are displayed in Figure 4.1.

Mechanical Engineering: A mechanical engineer for Sand Semiconductor was given the task of finding another mold compound supplier that would give Sand's line of plastic packaged transistors better thermal properties. There are only four possible suppliers, A, B, C, and D, that offer a product that meets Sand's general specification. The engineer ordered from each of the suppliers more than enough mold compound to perform the experiment. However, when ordering the transistor die from the wafer fab, she realized that the transistors would come from four different wafers. This told her that a randomized block design would be required. After sawing the wafers into die, she randomly separated the die from each wafer into four groups. Each group was then assembled and molded with each mold compound. The θjc was measured for each transistor and the average was determined for each group. The resulting data are displayed in Figure 4.1.

Process Engineering: A process engineer at Shippy's Sheet Plastics was interested in evaluating the life of four different saw-blade manufacturers, A, B, C, and D. Because the blades already last for several days, and the engineer wanted to evaluate the average of a minimum of four blades for each of the manufacturers, the experiment would take at least two months if run on the same machine. This was unacceptable to the engineer's supervisor, and the engineer was forced to find an alternative means for running the experiment. He chose to use four machines in parallel. Each manufacturer's blade was run on each machine in random order. The new, randomized block design was accepted by the supervisor because it allowed the experiment to be run in approximately one-fourth the time of the original design. After running the experiment, the averages were calculated for each manufacturer. The resulting data are displayed in Figure 4.1.

Software Engineering: A software engineer for Annoyance Alarm Clocks, Inc., was evaluating four voice-command software packages, A, B, C, and D, for their fully automated, voice-activated assembly line. The entire assembly line for a particular clock requires only one operator to run. The factory operates around the clock, so to speak, seven days a week, and requires four shifts. Hence, there are four operators that are certified to run this particular line. The engineer decided to evaluate the four voice-command software packages by having each operator ask the system to perform a series of 100 predetermined commands. Each operator completed this with all four software packages. The number of errors performed by the system was recorded. The engineer designed and evaluated this experiment as a randomized block design. The resulting data are displayed in Figure 4.1.

Manufacturing: A wafer fab was in the process of eliminating all of the tweezers used in handling silicon wafers and replacing them with vacuum wands. An engineer was asked to determine which vacuum wand tips should be purchased for the fab. The manufacturer of the wands has a selection of four different tips, A, B, C, and D, that could be used with their wands. The engineer chose to perform the evaluation with four randomly selected operators from the fab. The experimental design for this evaluation was a randomized block design. Each operator would use each wand to load and unload a specified number of the most fragile wafers processed in the fab. The number of broken wafers for each operator and each wand was recorded. The resulting data are displayed in Figure 4.1.

Administration: The law office of Brickel, Brickel, Brickel, and Jones was interested in purchasing the latest and greatest word processing software for each of the four secretaries in the office. Jones, being the odd man out, was asked to determine which software package should be purchased for the office. Four popular software packages, A, B, C, and D, were to be evaluated. Jones chose to use a randomized block design, with each of the four secretaries typing a lengthy letter using each of the software packages. The number of errors for each secretary was recorded. The resulting data are displayed in Figure 4.1.

Facilities: Chilled water is used for many cooling needs at Orifice Extrusion, Inc. The facility has four main manufacturing buildings, which are all serviced by one chilled-water system. Due to signs of corrosion of the chilled-water system, the facilities engineer decided to investigate the use of an additive to eliminate this potential problem. There are several additives proven to be effective for this problem; however, the engineer wished to find one that would have the least amount of effect on the temperature at each manufacturing building at the facility, or, if possible, lower the temperature of the chilled water at each building. In other words, the lower the water temperature, the better. The experiment was completed during the Christmas vacation shut-down. The engineer randomly changed out the water with a different additive and measured the temperature of the water at each building. The resulting data are displayed in Figure 4.1.

Finance: The CEO of a large textile corporation, Solids, Inc., was concerned about the amount of money being spent on travel taken by employees to the many different facilities within the corporation as well as to customers. He has received several letters from four different airlines, A, B, C, and D, concerning savings that could be made if he chose their airline as the principal carrier for his company. He asked his general manager and corporate vice president of finance to look into this issue, who immediately labeled it as a project for the new accountant that had been hired from the local university. Because she wanted to find the lowest overall cost for travel to the most popular locations, the accountant drew up a Pareto of the locations that were most frequently traveled to by her fellow employees. It was obvious to her that she needed to design a randomized block design. She then proceeded to call the different airlines for the discounted rates to the four locations most traveled by the company. The resulting data are displayed in Figure 4.1.

Personnel: A college recruiter for Asparagus Computer, Inc., was interested in determining which form of advertising he should use to attract graduate-level engineering students to his companies on campus interviews. His experiment would include advertisements in radio (A), school newspapers (B), bulletin board advertisements in the engineering buildings (C), and fliers to be distributed at the college bookstore (D). He also chose to run this experiment at the first four universities that he would interview at, and use the results for the remaining universities. His response was the number of graduate engineering students that came by for interviews. His experiment was a randomized block design. The resulting data are displayed in Figure 4.1.

Sales: A sales manager for Terrific Toys, Inc., was interested in determining which of the many different restaurants in the metropolitan area was the best for dining and wining customers. After discussing this with her three salesmen, it was obvious that there were four restaurants that stood above the rest. She decided that she and her salesmen would critique the four contenders with a survey that they had generated while playing their weekly Wednesday golf match. The experimental design that they created was a randomized block design. They proceeded to have dinner at each of the restaurants and independently scored their surveys. The resulting data are displayed in Figure 4.1.

7   APPLICATION ROADMAP

Step 1. State the null and alternate hypothesis:

$H0$: $\mu_1 = \mu_2 = \mu_3 = \mu_4$ or all the treatment
means are equal.

$H1$: At least two of the treatment means
are not equal.

Treatments	Blocks				T_i
	A	B	C	D	
1	25	33	36	23	117
2	23	22	11	17	73
3	9	6	3	12	30
4	19	20	22	29	90
$T_{.j}$	76	81	72	81	$T_{..} = 310$

FIGURE 4.2.   Coded Scenario Data with Factor and Block Totals

Step 2. State the risk level:

$\alpha = 0.05$, or equivalently $\alpha = 5\%$.

Step 3. Computations.

Step 3.1. Compute the totals for each treatment (T_i) and record this number to the right of the corresponding treatment row:

$T_{1.} = 25 + 33 + 36 + 23 = 117.$

$T_{2.} = 23 + 22 + 11 + 17 = 73.$

$T_{3.} = 9 + 6 + 3 + 12 = 30.$

$T_{4.} = 19 + 20 + 22 + 29 = 90.$

Step 3.2. Compute the totals for each block ($T_{.j}$) and record this number at the bottom of the corresponding block column:

$T_{.1} = 25 + 23 + 9 + 19 = 76.$

$T_{.2} = 33 + 22 + 6 + 20 = 81.$

$T_{.3} = 36 + 11 + 3 + 22 = 72.$

$T_{.4} = 23 + 17 + 12 + 29 = 81.$

Step 3.3. Compute the grand total for all the treatments ($T_{..}$) and record at the bottom of the treatment totals column:

$T_{..} = 117 + 73 + 30 + 90 = 310.$

Step 3.4. Construct the ANOVA table as shown in Table 4.1. Replace the entries in the table with the computed values.

Step 3.5. Compute the degrees of freedom for treatments and record in the ANOVA table:

$$\text{degrees of freedom (treatments)}$$
$$= \text{number of treatments} - 1 = a - 1,$$
$$\text{degrees of freedom (treatments)}$$
$$= 4 - 1 = 3.$$

Step 3.6. Compute the degrees of freedom for blocks and record in the ANOVA table:

$$\text{degrees of freedom (blocks)}$$
$$= \text{number of blocks} - 1 = b - 1,$$
$$\text{degrees of freedom (blocks)}$$
$$= 4 - 1 = 3.$$

Step 3.7. Compute the degrees of freedom for error and record in the ANOVA table:

$$\text{degrees of freedom (error)} = (a-1)(b-1),$$
$$\text{degrees of freedom (error)} = (4-1)(4-1)$$
$$= 3 \times 3 = 9.$$

TABLE 4.1   *General Analysis of Variance Table*

Source of Variation	Degrees of Freedom	Sum of Squares	Mean Square	F
A Treatments	$a - 1$	SSA	$MSA = \dfrac{SSA}{a-1}$	$\dfrac{MSA}{MSE}$
B Blocks	$b - 1$	SSB	$MSB = \dfrac{SSB}{b-1}$	
Error	$(a-1)(b-1)$	SSE	$MSE = \dfrac{SSE}{(a-1)(b-1)}$	
Total	$ab - 1$	SST		

Step 3.8. Compute the degrees of freedom for the total and record in the ANOVA table. Add the degrees of freedom for the treatments, blocks, and error. This sum should equal the degrees of freedom for the total.

$$\text{degrees of freedom (total)}$$
$$= \text{total number of observations}$$
$$- 1 = ab - 1,$$
$$\text{degrees of freedom (total)}$$
$$= (4 \times 4) - 1 = 16 - 1 = 15.$$

Step 3.9. Compute the correction factor:

$$C = \frac{T_{..}^2}{ab} = \frac{96100}{16} = 6006.25.$$

Step 3.10. Compute the sum of squares total (SST) and record in the ANOVA table:

$$SST = \sum_{i=1}^{a} \sum_{j=1}^{b} y_{ij}^2 - C,$$

$$SST = (25^2 + 23^2 + 9^2 + 19^2 + 33^2$$
$$+ 22^2 + 6^2 + 20^2 + 36^2 + 11^2$$
$$+ 3^2 + 22^2 + 23^2 + 17^2$$
$$+ 12^2 + 29^2) - C,$$
$$SST = 7318 - 6006.25 = 1311.75.$$

Step 3.11. Compute the sum of squares for the treatments (SSA) and record in the ANOVA table:

$$SSA = \sum_{i=1}^{a} \frac{T_{i.}^2}{b} - C,$$

$$SSA = \frac{(117^2 + 73^2 + 30^2 + 90^2)}{4} - C,$$

$$SSA = \frac{28018}{4} - 6006.25 = 7004.5$$
$$- 6006.25 = 998.25.$$

Step 3.12. Compute the sum of squares for the blocks (SSB) and record in the ANOVA table:

$$SSB = \sum_{j=1}^{b} \frac{T_{\cdot j}^2}{a} - C,$$

$$SSB = \frac{(76^2 + 81^2 + 72^2 + 81^2)}{4} - C,$$

$$SSB = \frac{24082}{4} - 6006.25 = 6020.5$$

$$- 6006.25 = 14.25.$$

Step 3.13. Compute the sum of squares for the error (SSE) and record in the ANOVA table:

$$SSE = SST - SSA - SSB = 1311.75 - 998.25$$

$$- 14.25 = 299.25.$$

Step 3.14. Compute the mean square treatments (MSA) and record in the ANOVA table:

$$MSA = \frac{SSA}{\text{degrees of freedom for treatments}}$$

$$= \frac{998.25}{3} = 332.75.$$

Step 3.15. Compute the mean square blocks (MSB) and record in the ANOVA table:

$$MSB = \frac{SSB}{\text{degrees of freedom for blocks}}$$

$$= \frac{14.25}{3} = 4.75.$$

Step 3.16. Compute the mean square error (MSE) and record in the ANOVA table:

$$MSE = \frac{SSE}{\text{degrees of freedom for error}}$$

$$= \frac{299.25}{9} = 33.25.$$

Step 3.17. Compute the F-value for treatments and record in the ANOVA table. After this step, the ANOVA table should look like Table 4.2.

$$F = \frac{MSA}{MSE} = \frac{332.75}{33.25} = 10.0075.$$

Step 4. Determine the critical F-region for the A treatments. Look up F_{α,ν_1,ν_2} in an F-table; where α is the risk factor, ν_1 is the number of degrees of freedom for the A treatments, and ν_2 is the number of degrees of freedom for the error. F-tables can be found in all the references as well as most statistics books.

$$F_{\alpha,\nu_1,\nu_2} = F_{0.05,3,9} = 3.86.$$

Step 5. Draw conclusions. Since the calculated F-value, 10.0075, is greater than the critical F-value, 3.86, we conclude that at least one of the treatment means is different than the others. The analysis of variance does not tell us which means are different. There are several methods to determine which means differ. Some of these techniques are Duncan's Multiple Range Test, the Least Significant Difference (LSD) Method, Newman-Keul's Test, and Tukey's Test. Descriptions of these techniques can be found in a variety of statistics books.

TABLE 4.2 *Completed Analysis of Variance Table*

Source of Variation	Degrees of Freedom	Sum of Squares	Mean Square	F
A Treatments	3	998.25	332.75	10.0075
B Blocks	3	14.25	4.75	
Error	9	299.25	33.25	
Total	15	1311.75		

REFERENCES AND BIBLIOGRAPHY

Hicks, C. R. 1964. *Fundamental Concepts in the Design of Experiments*. New York: Holt, Rinehart & Winston.

Hines, W. W., and D. C. Montgomery. 1991. *Probability and Statistics in Engineering and Management Science*, 3rd Edition. New York: John Wiley and Sons.

Miller, I., and J. E. Freund. 1985. *Probability and Statistics for Engineers*, 3rd Edition. Englewood Cliffs, New Jersey: Prentice-Hall.

Montgomery, D. C. 1991. *Design and Analysis of Experiments*, 3rd Edition. New York: John Wiley and Sons.

Walpole, R. E., and R. H. Myers. 1989. *Probability and Statistics for Engineers and Scientists*, 4th Edition. New York: MacMillan.

Index